50 Projects for Radio Amateurs

Edited By

Mike Browne, G3DIH

Radio Society of Great Britain

Published by the Radio Society of Great Britain, 3 Abbey Court, Fraser Road, Priory Business Park, Bedford MK44 3WH. Tel: +44 (0) 1234 832700 www.rsgb.org

First Printed 2018

Digitally printed 2020 onwards

Layout and Edited by: Mike Browne, G3DIH

Cover Design: Kevin Williams, M6CYB

Production: Mark Allgar, M1MPA

Printed in Great Britain by 4Edge Ltd of Hockley, Essex

ISBN: 9781 9101 9352 5

Contents

Preface

This book is a compilation of projects which appeared in publication such as 'Radio & Electronics Cookbook', 'Practical Projects' and 'Weekend Projects' all published a few years ago by the RSGB. They proved very popular both with newcomers to the hobby and with 'old hands' who wanted to try something not too difficult. It also provided everyone who read it with a source of ideas for other projects, which is what amateur radio is all about. There are also projects from issues of RadCom and RadCom Plus. Some of the projects are tried and tested projects and fairly easy to complete.

In this book you will find projects from a simple 'paddle key' and 'portable antenna' to more advanced projects such as a '80m transceiver' and '70cm handheld'.

Although I have made every effort to check that components are still available through RS components and Maplin etc. there remains the ever-present problem of sourcing those components which disappear from the shelves of stockists between the submission of the articles and the appearance of a book such as this. Radio amateurs however have over the years, developed the instincts of the magpie, and can be seen at rallies poring over junk to see if that elusive chip or transistor can be found for their latest project. If that component still eludes you, why not ask members of your local radio club? They will be pleased to help in any way they can. If you don't know where your nearest club is, you can find out information of clubs in your area by looking on the RSGBs website or in the RSGB Yearbook.

I acknowledge all previous contributors for their articles that appear in this book.

Mike Browne, G3DIH

January 2018

An Audio Filter

This self-contained and inexpensive audio filter fits in the lead between a receiver and the headphones. Active audio filter circuits have been published in amateur radio journals over many years. Some are very elaborate and seem to have as many components as the receiver itself. Some are very clever, having several controls for adjusting the bandwidth and the frequency response, notch or peak tuning, and other functions. These tend to invite the operator to spend more time testing and driving the filter than operating the receiver! Read on and find out why this one is different.

ABOUT THE FILTER

This audio filter unit is elementary in concept and has only one control – a switch to turn it on and off and at the same time switch the filter in and out of circuit. Thus, instant checks can be made on the effectiveness of the unit. In these days of DSP, the inexpensive, simple, single-integrated-circuit audio filter still has a useful place in amateur radio. It can perhaps best be described as an *audio pass-band modifier*.

The filter is a very effective and useful accessory for CW and SSB reception. To keep it simple and to avoid adjustable controls, a 'most-often-used' response characteristic has been adopted, with a peak around 800Hz. This frequency was chosen principally for CW reception but is also usefully placed and shaped for SSB reception. It attenuates both the low and the high audio frequencies, narrowing the pass-band.

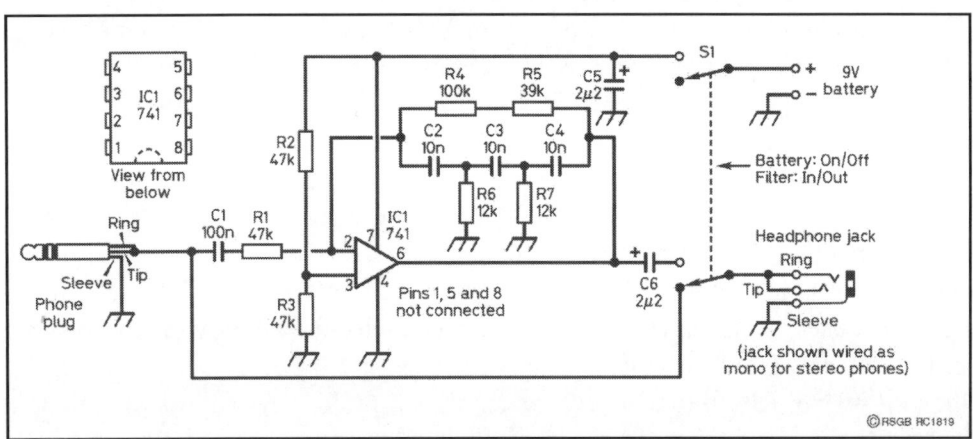

Fig 1. *The simple audio filter. Note the pin assignment of IC1 shows the IC as it would be seen when wired-up 'dead bug' style.*

This narrower pass-band reduces the unwanted noise heard in the headphones. However, this filter cannot eliminate or reduce the image sideband from the direct conversion process used in simple receivers.

THE CIRCUIT

The active element of the circuit, shown in **Fig 1**, is an operational amplifier integrated circuit, using the common, inexpensive, general-purpose LM741, with a bridged RC filter connected in its feedback path. How it works is a very interesting academic study which need not concern us here. The RC filter comprises C2, C3 and C4, along with R4, R5, R6 and R7. The values of these filter components are important. Use good-quality components, preferably of identical type. The normal tolerance spread of component values results in a flat 'peak' in the response characteristic, making it acceptable for SSB and CW reception. Several filters have been built using store-bought components with the values shown. No special selection seems necessary. A 9V battery powers the unit. Remember to observe polarity and to switch it off when not in use.

CONSTRUCTION

The whole device fits into a plastic box. Select your own type! The double-pole changeover toggle switch is mounted on the lid of the box. The IC is glued upside down to a piece of printed circuit board which is bolted to the floor of the box. Wiring is by the 'dead bug' method. Note that IC pin numbers 1, 5 and 8 are not connected. Two pieces of tag strip hold the RC filter components and support other wiring junctions. The battery is held in place in the box by double-sided sticky tape.

There is nothing critical about the construction, and instability was not experienced. The whole unit can be built in an evening. It has a high 'go-first-time' factor. There is nothing that needs any setting-up but feel free to experiment!

TESTING

To plot a filter frequency response you need access to quality laboratory instruments with known specifications. Trial tests using lesser-grade 'test meters' have been found to be prone to measurement errors, due in particular to insufficient information being known about the frequency characteristics of the measuring apparatus used. Consequently, no simple procedure for testing with cheaper test gear is offered. However, it is easy and convincing to check the effectiveness of the unit by ear. Listen to a strong carrier signal, preferably with no modulation. Tune to what you estimate to be a typical signal tone or note for CW reception – about 800Hz.

Switch the filter in and out of circuit. Adjust the pitch of the received signal very carefully. You should find a position where the amplitude (loudness) of the wanted signal rises when the filter is in circuit. Ignore any change in the noise level, just listen to and refer to the change in the level of the received signal alone, as the filter is switched in and out of circuit. Above and below this frequency you should find that the loudness of the signal is higher when the filter is out of circuit, confirming a peak in the overall filter response. There is a very noticeable reduction in noise when the filter is in circuit.

PARTS LIST

Resistors –all resistors $\frac{1}{2}$W metal film

R1 – 3 47k, 5%

R4 100k, 1%

R5 39k, 1%

R6, 7 12k, 1%

Capacitors

C1 100n

C2 – 4 10n, 1%

C5, 6 2.2µ, 16V electrolytic

Semiconductor

IC1 LM741

Additional items

S1 DPDT toggle

Battery clip

PP3 battery

Case

Audio connectors (as appropriate)

A Direct-Reading Capacitance Meter

Bags of surplus unused capacitors from rallies become an attractive proposition provided you can fathom out what they are. You may also possess polystyrene capacitors where the marking ink (used to show the values and working voltage) has rubbed off. This instrument saves you trying to recall myriads of capacitor value codings. You will find other uses, for example the ability to check the setting obtained with a variable capacitor used to find a resonance. The reading obtained by this direct-reading capacitance meter gives the value needed for a substitute fixed capacitor.

THE CHOICE

There have been various approaches made in designing capacitance checkers, including:

Multi-range test meter

A suitably-scaled multimeter can give approximate values for capacitors, using what is really one of the AC voltage test ranges in conjunction with a mains-powered step-down transformer. Such a solution is messy but reasonably effective. It is not a recommended practice with low-voltage capacitors or for use by beginners!

Test meter with internal AC source

Many modern digital multimeters have capacitance ranges. An alternating voltage is generated by an oscillator powered from internal batteries. These meters have their uses, though, but are expensive. However, the display format provided, which usually consists of three-and-a-half digits, is unsuitable for taking variable readings, as when checking a suspect plastic dielectric tuning capacitor.

Service technician's resistance / capacitance test bridge

Resistance/capacitance ratio bridges used to be one of the tools of every service technician. These versatile units often have additional features. There is usually a source of high DC voltage (0 to 500V variable) for the forming of high-voltage electrolytic capacitors. This voltage may also used for testing leaking paper capacitors which, as every service technician knows, contribute to the demise of resistors followed by valves and transformers.

Purpose-designed capacitance checker

A direct-reading capacitance meter has the facility to transfer capacitive reactance, in linear fashion, to the scale of an analogue direct-reading meter. The analogue meter is equally happy with fixed or variable capacitance readings.

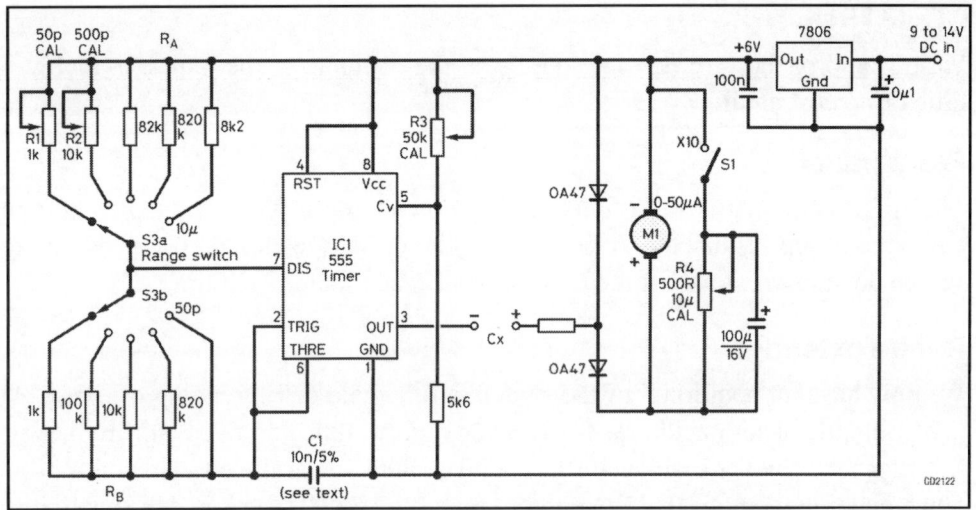

Fig 1. *Circuit diagram of the capacitance meter. The resistor groups RA and RB are discussed in the text, as are the pair of resistors selected by switch S3. S3a selects RA, and RB is selected by S3b. All fixed resistors are 5% ¼W carbon types, and polarised capacitors are electrolytics.*

The design of cheap and accurate direct-reading capacitance meters becomes a lot easier with the availability of solid-state integrated circuits. The simplest designs settled on the use of the 555 timer IC operated as a monostable oscillator.

The definitive work was done by A Wilcox for the magazine *Television* in the early 1970s. Constructional projects subsequently appeared in *Electronics Australia* October 1976, and in *QST* January 1983. The design used for this project begins with the *QST* circuit and then sets out to overcome many of the perceived limitations.

THEORY

An explanation is provided for those who like to know a little more about such things. To make the theory aspects clearer, please refer to the circuit given in **Fig 1**.

The 555 is used as a monostable oscillator and the test capacitor, which we will call Cx, is first charged and then discharged, and the meter indicates the average discharge current. The formula is:

$$I_{ave} = \frac{V\ C_x}{(R_A + 2R_B)\ C_1} K$$

where V is the voltage to which Cx is charged and K is a constant, depending on the charge and discharge time of the 555 circuit, including the contribution made by the IC internal structure as well as the external resistance ratios and C1.

5

CONSTRUCTION

Various approaches are possible but using the contents of the junk box involves little or no cost at all.

Front panel

The specification calls for a 50µA meter mounted on a plastic box. The bottom range reads 50pF full scale. A switch provides extra ranges. There are five rotary switch positions, extending the ranges by a decade factor each time.

Range extender

We now have, at position 5 of the switch, a full-scale direct reading of 0.5µF. At this point, the meter needle starts to flicker visibly at the 555 oscillator frequency. In order to extend the range further, a ×10 shunt is paralleled across the basic 50µA meter movement. This extends all ranges by a ×10 factor, and the maximum becomes 5µF. As the meter flicker is unacceptable, an auxiliary switch also brings in a 100µF smoothing capacitor across the meter movement.

The applied voltage is about 2V on the terminals, which is unlikely to harm even the most delicate components.

Construction details

Short internal wiring is preferable and the IC PCB is mounted right at the test terminals. Additionally, the 555 is mounted in a socket for peace of mind, allowing easy replacement should this ever be necessary.

Calibration

Use a good 0.047µF capacitor on range 3 to set the 50kW calibrator to produce a reading of 47µA. Then set the ×10 range (using the appropriate switch settings) with its preset to read 5µF full scale. Next adjust the 50pF and 500pF controls. You are now able to see the difference between 2.7pF and 3.3pF accurately.

Low-cost operation

The unit has a small mains power supply and needs a secondary winding able to provide anything between 9V and 14V DC to the regulator input. If you do not like to wire mains-operated accessories, a 9V alkaline battery can be used, but do not omit the voltage regulator.

CONCLUSION

There is great satisfaction to be had in building up reliable and accurate test equipment. Spend some time on the cosmetics; try to obtain matching knobs, give consideration to using brass flathead machine screws where they appear on the outside, and if silk-screening is out of the question you can use Dymotape® labels. Rubber stick-on feet ensure the case does not scratch anything and stops it sliding off the bench.

With a few known 2% calibration capacitors used from time to time as reference sources, this meter has been found to provide consistent and quite accurate measurements. The prototype has been in regular use for several months and performs consistently well.

PARTS LIST

Numbered components not listed below are for text reference only.

Resistors

R1 1k PC-mounting
R2 10k PC-mounting
R3 50k PC-mounting

Semiconductor

IC1   555 timer
S1 SPST switch

Hardware

S3 2-pole 5-way rotary switch

Field-Strength Measurement

Never expect the same antenna to work the same way in two locations. Ground conductivity and reflections from objects, even at considerable distances, can upset our expectations. The effects are more pronounced on VHF, and experimentation with a field-strength indicator (not 'meter', because no absolute accuracy can be claimed) can be extremely enlightening.

USE OF THE INDICATOR

You can get some idea of the field-strength pattern of your antenna by taking a reading on your completed field-strength indicator, then walking around and noting how the reading changes.

CONSTRUCTION

This indicator was originally designed for a specific use: testing glider radio systems. Because the indicator has subsequently proved so useful for all amateur radio bands from 160m to 70cm, it is worth passing on.

Several designs were tried to begin with, the most interesting being a dipole with a 68W resistor connected between the two halves and an RF detector connected across the resistor. With the dipole securely mounted on a tripod, the dipole length could be adjusted for maximum reading, enabling the wavelength of the signal to be measured with surprising accuracy.

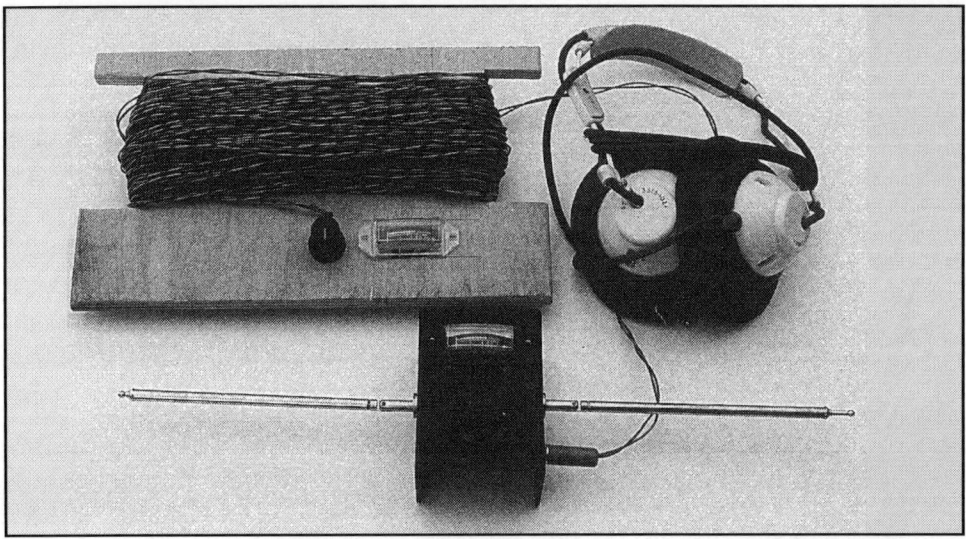

The completed indicator ready for use

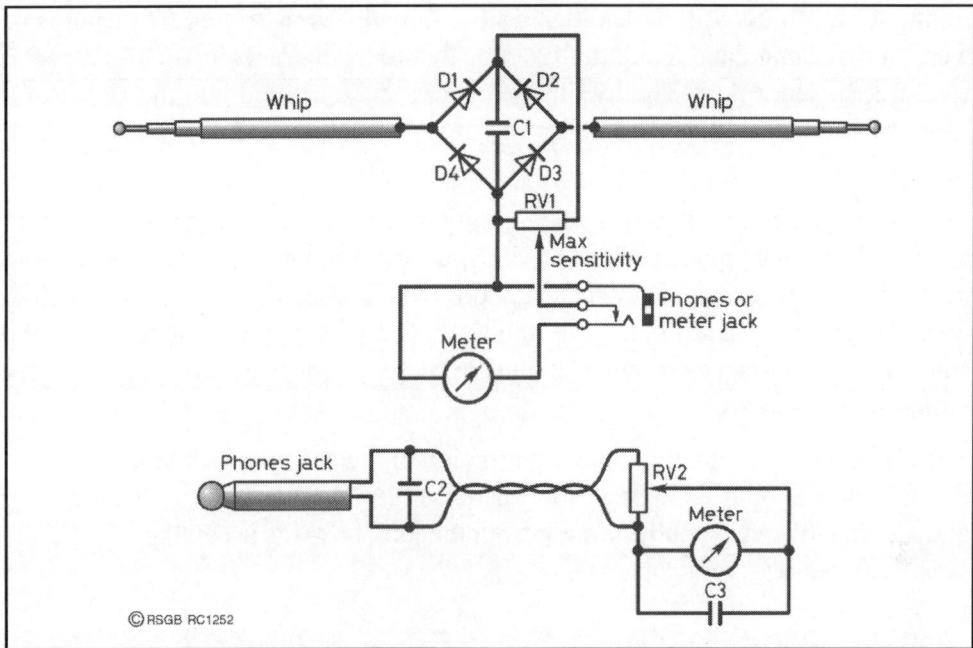

Fig 1: *Circuit diagram of the field strength indicator*

Unfortunately, the sensitivity of the unit was not sufficient for the purpose, so the resistor and detector were replaced by a full-wave bridge as shown in **Fig 1**.

A wide-band amplifier could have been added to the original circuit to gain the necessary sensitivity, but experience has taught me that it is far better to use a 'passive' unit than one requiring a power source, as the battery has the annoying habit of being flat when needed most urgently.

In the circuit diagram there is a jack socket labelled 'phones'. This has two functions. On the aircraft band, amplitude modulation is used, and the modulation can be checked by plugging a pair of high-impedance headphones into this socket. Unfortunately, FM is the most popular mode used on the amateur VHF bands and this is not rendered audible using a diode detector.

The second function enables an extension meter to be used. This consists of a second meter and sensitivity control which, when used, requires the sensitivity control in the head unit to be set to maximum. The extension cable can either be two lengths of hook-up wire twisted together or a length of coax. If you have a length of coax in your junk box which has been removed from service as it has become lossy at RF, it would suit this job admirably because a DC measurements are unaffected by the problems that cause the cable to be lossy at VHF.

A simple way to twist long lengths of hook-up wire is to lay out two lengths of wire on the ground and make the far ends fast to a solid anchor such as a bench vice, tree or fence post. The two free ends are then clamped into the chuck of a hand drill and, with the wires under light tension, the drill is used to twist them together.

The SWR bridge is very useful when tuning matching networks at the base of an antenna, but can be misleading at times. Spurious resonance within the matching network can, on occasions, indicate a good VSWR when there is a poor match to the antenna. Using a field-strength indicator on a tripod some distance from the antenna and an extension meter alongside the operator enables quick and easy tuning-up of systems.

There is little to be said about the construction of the indicator, with the exception that all wires should be kept as short as possible. The meter recommended is a 50μA unit. However, a 200μA meter from the junk box was used; this reduced the sensitivity but was very cheap!

PARTS LIST

RV1, 2	10k linear potentiometer
C1–3	10n ceramic
D1–4	OA2 germanium diode or similar
Meter	50μA FSD
Antenna halves	560mm telescopic whips
Jack socket with switch and plug to match	

A Bi-Directional Wattmeter

Three members of a radio club are building 80m CW transceivers for holiday use. A power / SWR meter was needed that could operate at low power and did not need to be adjusted every time the frequency or power was changed. This is the design that was chosen.

HOW IT WORKS

When the transmitter is connected to P1 (**Fig 1**) and the aerial to P2, 99% of the transmitter power arrives at P2 to go to the aerial. The other 1% is sampled by the transformers and fed to R1 and R2. The RF voltage across R1 / R2 is rectified by D1 and fed to meter M1. The phasing of the windings on transformers T1 and T2 cancels any voltage on R3 and R4.

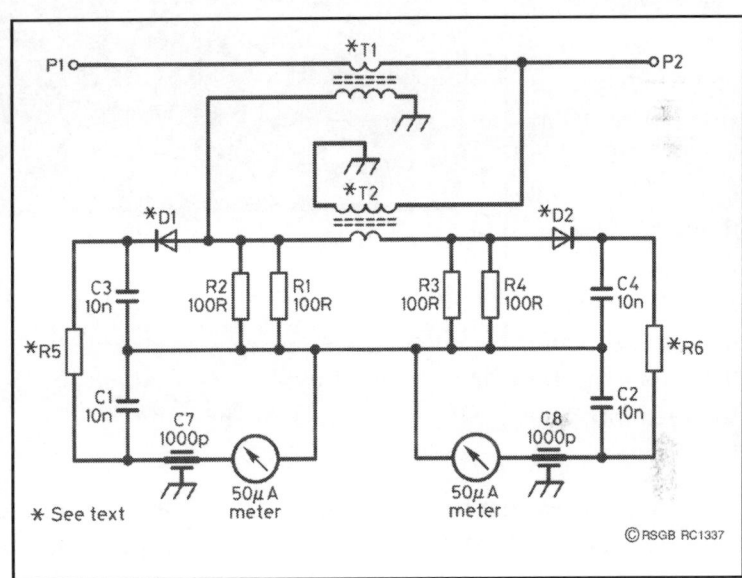

If the aerial does not present a perfect 50W impedance, some power will be reflected. This re-

Fig 1. Circuit diagram of wattmeter.

flected power is 180° out of phase with the transmitted power. About 1% of this reflected power is sampled by the transformers and fed to R3 and R4. The RF voltage across R3 and R4 is rectified by D2 and fed to meter M2. The phasing of the windings on transformers T1 and T2 cancel any voltage on R1 and R2.

CONSTRUCTION

The power meter RF components must be constructed in a metal box (or a box made from PCB material), as shown in **Fig 2**. The transformers are wound on ferrite beads (Fair-Rite 26-43006302, Bonex 6302 PA balun bead), which are fairly large and are easy to handle.

Each transformer has 14 turns of 26SWG enamelled copper wire wound tightly on the core with the turns equally spaced. Ensure that both transformers are wound with the same number of turns. The second winding consists of a short length of thin coax threaded through the centre, one end of its braid connected to ground, and the other neatly trimmed off and left disconnected.

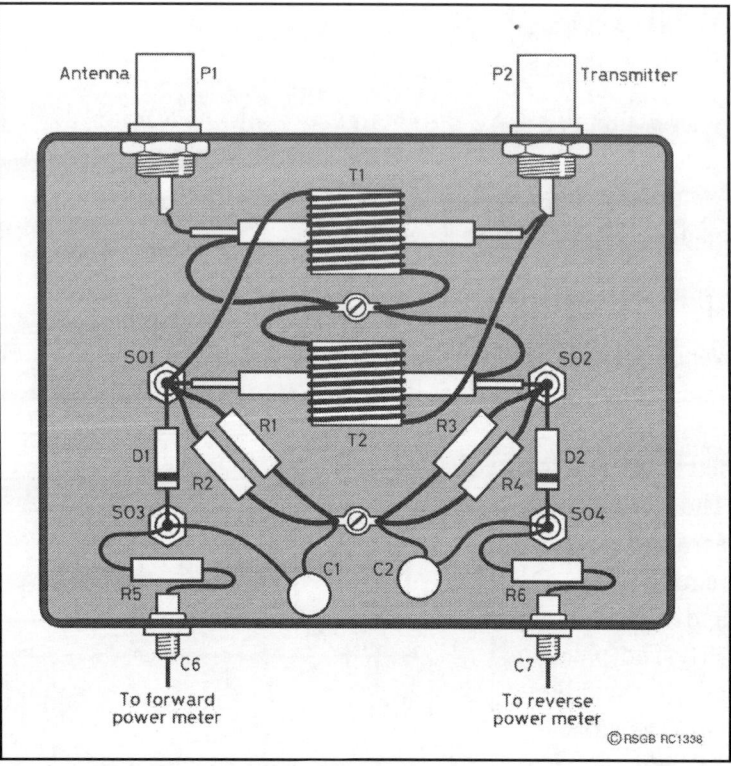

Fig 2. Component layout (meters not shown).

The braid screens the two windings as far as the electrostatic field is concerned but will not hinder the electromagnetic field. (This is known as a Faraday cage or screen).

If 'junk box' meters are used, it will be necessary to calibrate the unit using a known power meter. This is not difficult to do, provided your transmitter can be reduced progressively in output to about 1W. This is the procedure to follow.

CALIBRATION

Connect the transmitter output to your power meter, the output of which must then be connected to the reference power meter which in turn must be connected to a 50W dummy load. A 100kW variable resistor is connected in place of R5 and the transmitter adjusted to give an output of 10W, as indicated on the reference meter. The 100kW pot is then adjusted for full-scale deflection (FSD), its value is measured and R5 and R6 replaced by resistors of the correct values. By progressively reducing the output power a graph of calibration can be made for the meter and from that a new meter scale can be laid out.

Making scales is easier than most people realise, provided a photocopier with a zoom facility is available. Remove the old scale and photocopy it onto paper at four times its original size. On the photocopy, the new scale is then carefully marked over the old scale using Letraset figures and lines, and all unwanted markings are covered with Tipp-Ex. Now, rather than reducing the size by four on the photocopier, reduce it by two and then touch out any remaining imperfections using Tipp-Ex, then reduce by two again to regain the original size. Using either spray-on adhesive or thin double-sided tape the new scale is stuck over the old scale and re-mounted in the meter casing. This power meter is designed for 10W, but the 43 mix beads can be used up to 100W without problems. Calibration is carried out the same way but it is necessary to increase the power rating of R1, 2, 3 and 4. Two 100W 1/8W resistors were used to give 50W at 1/4W for 10W. For the 100W test, four 100W resistors were connected in series / parallel to give 50W at 1/2W. On SSB they do get warm, so if SSTV or very-high-duty-cycle modes are to be used, it may be advisable to increase the power rating further.

PARTS LIST

Resistors

R1–4	100R carbon or metal film

Capacitors

C1–4	10n disc ceramic
C7, 8	1000p feedthrough

Additional items

T1, 2	14 turns of 26SWG copper wire on a Fair-Rite 26-43006302 ferrite ring

A Switched Attenuator

How many times has a signal been too strong for the experiment you wish to carry out? It could be from an oscillator on the bench or from signals from an aerial overpowering a mixer. This attenuator will solve those problems.

WHERE CAN I USE IT?

Applications other than those given above are if you are interested in DF and need to attenuate the signal when getting close to the transmitter, or if you have a problem with a TV signal being too strong and causing ghosting on other signals (cross-modulation).

ATTENUATORS

These problems can be averted by using a switched attenuator (or 'pad'). The times that we need attenuators occur far more often than first realised. When designing the attenuators, account must be taken of the distinct possibility of poor screening. There is hardly any point in designing a 20dB attenuator when the leakage around the circuit is approaching this value. It is also important to decide on the accuracy required. If it is intended to do very accurate measurements, the construction has to be impeccable, but for comparisons between signals it would be possible to accept attenuation values to a lower degree of accuracy. [A 'pad' is the name given to a group of components with a known attenuation. – Ed.]

The most useful attenuator is a switched unit where a range from zero to over 60dB in 1dB steps can be covered. This is not as difficult as it first seems because,

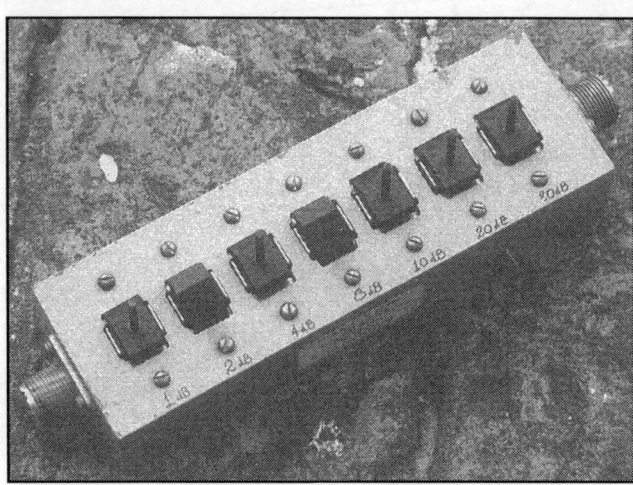

The completed attenuator.

by summing different attenuators, we can obtain the value we need. It takes only seven switches to cover 65dB. The seven values of attenuation are 1dB, 2dB, 4dB, 8dB, 10dB, and two at 20dB; these can be switched in or out at will. As an example, if 47dB were needed, switch on the two 20dB pads plus the 4, 2 and 1dB pads.

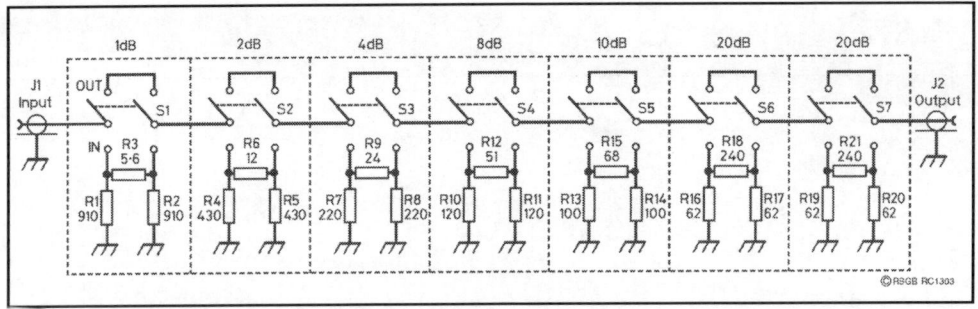

Fig 1. The attenuator consists of seven p-network sections, so-called because each pad (eg R1, R3 and R2) resemble the Greek letter pi (p). Input and output impedances are 50W.

CONSTRUCTION

The prototype attenuator is shown in the photo and in **Fig 1**. It is housed in a box made from epoxy PCB material. The top and sides are cut to size and soldered into a box shape.

It is easier to cut the switch holes prior to making the box. After the box has been constructed, screens made from thin brass shim should be cut and soldered between the switch holes. Next, the switches are fitted and the unit wired up. When this is done, the unit is checked and a back cover, securely earthed to the box, is fitted.

COMPONENTS

The switches must have low capacitance between the contacts and simple slide switches are the best selection. The Maplin DPDT miniature switch (FH36P) would be suitable and Maplin also supplies 1% resistors. Connectors to the unit must be coaxial but can be left to personal preference. The resistor values shown in **Fig 1** determine the attenuator's accuracy at around 5%. This is done for practical reasons. For example, if we wanted to make the attenuation value of the 4dB cell exactly 4dB, the resistor values would have to be 220.97W and 23.85W. You will see from **Fig 1** that the values used are 220W and 24W, giving an attenuation value of 4.02dB.

PARTS LIST	
Resistors All resistors in ohms, 1/2W or 1W carbon or metal film	
R1, 2	910
R3	5.6
R4, 5	430

15

PARTS LIST (Contd)

R6	12
R7, 8	220
R9	24
R10, 11	120
R12	51
R13, 14	100
R15	68
R16, 17, 19, 20	62
R18, 21	240

Other items

S1 – 7 double pole miniature switch

J1, 2 SO239 sockets (or similar to suit your equipment)

A Signal Injector

One of the most satisfying things in amateur radio is being able to locate and repair faults on equipment. One of the most useful tools for fault detection is a signal generator but, as these are often quite bulky and expensive, here is a simple, low-cost alternative.

DESCRIPTION

The circuit shown in **Fig 1** is based around a two-transistor multivibrator, designed to produce square-wave oscillations at about 1kHz. Harmonics (at 3kHz, 5kHz, etc) ensure that the signal can be used not just to test audio circuits but RF circuits as well. When listening to the signal, it can best be described as a rather unpleasant buzz.

The multivibrator consists of a two-transistor circuit in which the transistors are alternately turned on and off. D1 and D2 are used to ensure the multivibrator produces square waves with very 'sharp' edges, and hence the greatest harmonic content. In theory, the harmonics from this circuit continue to infinity but, in practice, there is a limit at which they can be detected. With the circuit shown the harmonics are detectable beyond 145MHz.

Output from the oscillator is coupled to an amplifier (TR3) via C2. The amplifier is used to ensure that any loading imposed by the circuit under test will not cause the multivibrator to stop. Biasing resistor R5 is low in value compared to many

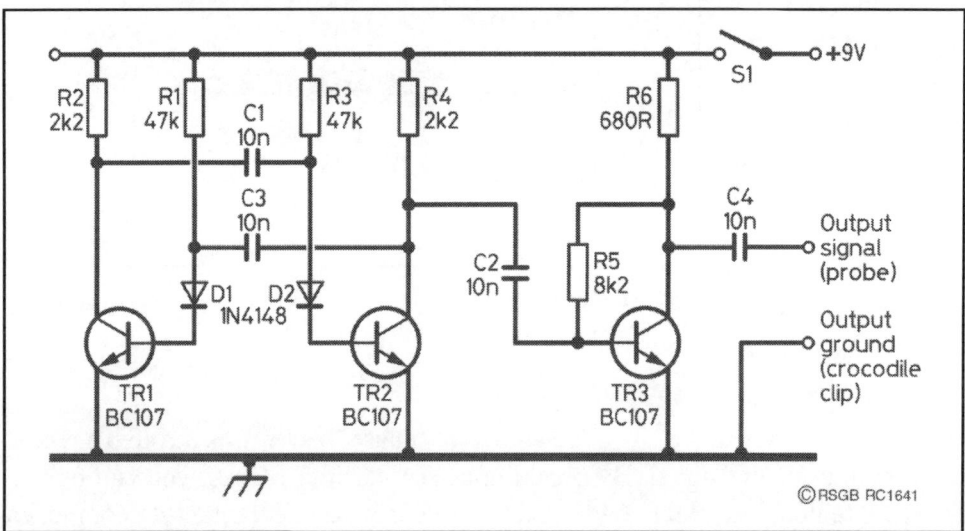

Fig 1. The simple signal injector is based on a multivibrator. This generates square waves, which are rich in odd harmonics.

audio amplifiers but this ensures that the harmonic content of the output is as high as possible. Output coupling is via C4, which must be rated at 50V minimum to provide isolation and protection from the circuit under test.

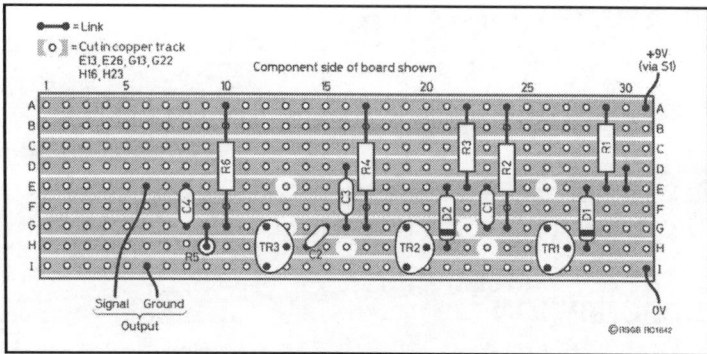

Fig 2. Veroboard layout of the simple signal injector.

CONSTRUCTION

The circuit is built on Veroboard (**Fig 2**), 9 strips × 30 holes. There are several track cuts, which can be made with a special track-cutting tool or a drill bit held in your hand. You can assemble the circuit in any order but, generally, the diodes and transistors are left to last. The output probe (**Fig 3**) is made from stiff copper wire, with the insulation left on for the majority of its length and the tip sharpened with a file to a point to ensure good contact. A lead with a crocodile clip is attached to the earth or chassis of the equipment under test.

TESTING AND USE OF THE SIGNAL INJECTOR

To test the circuit, touch the probe to either the input of an audio amplifier or the aerial socket of a receiver. If everything is working, you will hear a buzzing sound.

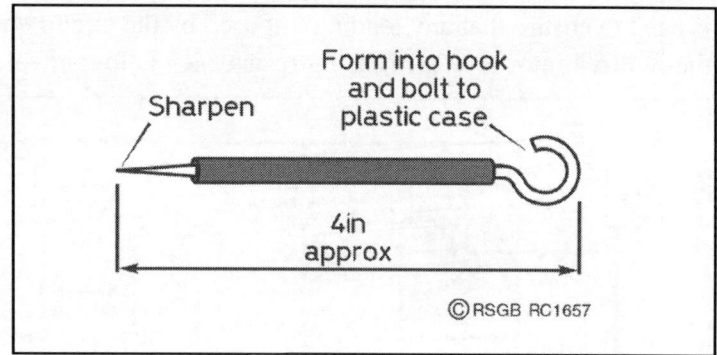

Fig 3. The output probe is made from stiff copper wire. Electrical connection is via a solder tag.

The basic idea of using a signal injector is to 'chase' backwards through a receiver, listening for the signal. By breaking a receiver into blocks, you can quickly isolate a fault to one block. Once located, you can pursue the fault to component level.

SAFETY NOTICE

The simple signal injector is designed to be connected to equipment which is powered on. Although there is a capacitor to provide protection from the voltages present in working circuits, the output must not be connected to any item of equipment which works on voltages higher than 24V. Television sets or any equipment which uses valves is unsuitable and unsafe for the use of this injector.

PARTS LIST

Resistors All resistors 1/4W, 10% tolerance, carbon or metal film

R1, 3	47k
R2, 4	2.2k
R5	8.2k
R6	680

Capacitors

C1–3	10n 16V (minimum) disc ceramic
C4	10n 50V (minimum) disc ceramic

Semiconductors

D1, 2	1N4148 or any silicon signal diode
TR1–3	BC107 / BC108 / BC109 or any similar npn small-signal type

Additional items

S1 On-off switch

Veroboard (stripboard), 0.1in pitch, 9 strips × 30 holes

PP3 battery clip

Hook-up wire

Crocodile clip

Plastic case to suit

Extending The Use Of Your Dip Oscillator

The grid dip oscillator (GDO) offers a quick and easy means of checking (to a degree of accuracy acceptable for experimental purposes) the inductance value of coils in the microhenry (μH) range and capacitors in the picofarad (pF) range, such as are commonly used in radio circuits. This can be very useful, for example, when constructing an ATU, a crystal set, a short-wave receiver, a VFO or a band-pass filter for a direct-conversion receiver.

DETERMINATION OF L AND/OR C

For this purpose, I keep with my GDO two fixed-value RF coils of known inductance – 4.7μH and 10μH – and one capacitor each of 47pF and 100pF (but the choice of values is yours). You may decide to keep one or more of each, to be selected from **Table 1** and **Table 2**.

My personal choice of coil type is the moulded RF choke (Maplin) or RF inductor (Mainline or RS). These are axial-lead, ferrite based, encapsulated, easy to handle, and readily available at low cost in a range of fixed values. The capacitors are 5% tolerance polystyrene, also axial-lead.

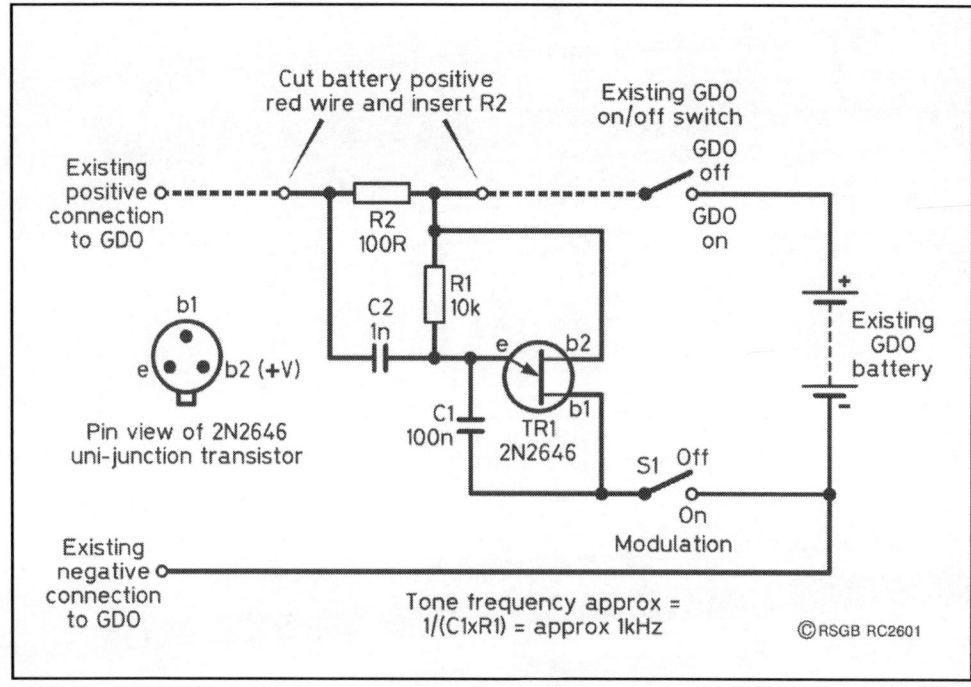

Fig 1. This audio oscillator adds a 1kHz AM tone to a GDO signal.

To determine or verify the value of either an RF coil or a capacitor, simply connect the unknown component in parallel with the appropriate known component to form a parallel LC tuned circuit, ie an unknown L in parallel with a known C (or vice versa), then use the GDO to determine the resonant frequency of the parallel LC circuit.

The value of the unknown component can then be obtained easily to an acceptable approximation, by using the relevant formula from Tables 1 and 2 and a pocket calculator.

Note that, in Tables 1 and 2, F is the frequency in megahertz as given by the GDO.

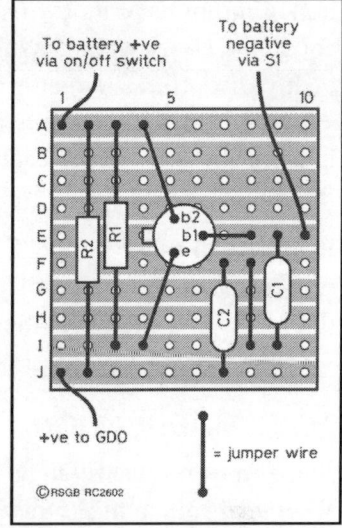

Fig 2. Stripboard layout.

Example 1

An unknown capacitor in parallel with my known 10μH inductor produces a dip at 6.1MHz, hence F = 6.1MHz. From **Table 1**, the value of the unknown capacitor is given by:

$$C = 2533 \div F2$$
$$= 2533 \div 6.1 \div 6.1$$
$$= 68pF.$$

Example 2

An unknown coil in parallel with my known 47pF capacitor produces a dip at 12.8MHz, hence F = 12.8MHz. From **Table 2**, the value of the unknown inductance is given by:

$$L = 539 \div F2$$
$$= 539 \div 12.8 \div 12.8$$
$$= 3.3μH.$$

Bear in mind that, because the accuracy of results relies upon the frequency as derived from the GDO, it would be sensible to keep the coupling between the LC circuit and the GDO as loose as possible, consistent with an observable dip. This minimises pulling of the GDO frequency. Also, rather than relying upon the frequency calibration of the GDO itself, it might be useful to monitor the GDO frequency on an HF receiver or a digital frequency meter.

A final point worth considering is that each fixed-value inductor of the type mentioned might have its own self-resonant frequency, but these would typically lie above the HF range so should not be a problem. For example, the self-resonance of my own 10μH inductor is about 50MHz and that of the 4.7μH one is about 70MHz. You could quickly and simply find out the self-resonant frequency of an inductor by taping it to each of the GDO coils in turn and tuning across the full frequency span.

It is best to make L and C measurements at frequencies much lower than the self-resonant frequency of your chosen test-inductor, but perhaps better be safe than sorry and stick with the lower microhenry values if your interest lies between 1.8 and 30MHz.

TONE MODULATION

Sometimes it is useful to be able to hear an audio tone when using the GDO as an RF signal source in association with a radio receiver.

If your GDO does not have tone modulation, you might like to construct the simple add-on 1kHz audio oscillator circuit shown in **Fig 1**. It uses a unijunction transistor, the frequency of oscillation being given approximately by 1/(R1 × C1). The 1kHz tone output connects via C2 to the positive supply line of the GDO, which it modulates.

R2 acts as the modulator load and its value helps to determine the level of modulation. This produces simple but effective tone modulation of the GDO's RF signal which can be heard on an AM or an FM receiver.

Fig 2 gives a suggested layout of the components on a small piece of stripboard, without the need for any track cutting. The finished board might conveniently mount on one of the GDO meter terminals, provided care is taken to isolate the copper-tracks from the terminal.

For known L (μH) of	1	2.2	4.7	6.8	10	22
Unknown C (pF)	$25330/F^2$	$11513/F^2$	$5389/F^2$	$3725/F^2$	$2533/F^2$	$1151/F^2$

Table 1. To determine an unknown capacitance

For known C (pF) of	10	22	33	47	68	100
Unknown L (μH)	$2533/F^2$	$1151/F^2$	$768/F^2$	$539/F^2$	$373/F^2$	$253/F^2$

Table 2. To determine an unknown inductance

PARTS LIST

Resistors	Capacitors	Semiconductor
R1 10k	C1 100n	TR1 2N2646
R2 100R	C2 1n	

A Diode/Transistor Tester   1.8

One of the regular problems when assembling diodes into a project is working out which end is the cathode and which the anode. This is particularly so when using diodes from the 'junk box' and finding that the markings have become somewhat indistinct. It is not too difficult to use a resistance meter to check the polarity but this only looks at one aspect. The unit described below provides an unambiguous means of identifying the cathode and carries out some other simple tests. It can also be used to check that bipolar transistors are 'in the land of the living'.

CAUTION

But first, a few words of caution: be very wary of using rectifiers of unknown characteristics in power supplies or other high-current applications.

The fact that a rectifier looks big enough is not a good indication that it is adequate for the job. Using inadequate devices is likely to result, at best, in poor reliability. At worst, the result could be the well-known 'dark brown smell', followed by smoke and even flames.

CIRCUIT DESCRIPTION

The circuit is shown in **Fig 1**. A simple square-wave oscillator is formed by inverters IC1a and IC1b. This runs at about 1.2kHz although the actual frequency is not particularly important.

Gates IC1c and IC1d buffer the output of the oscillator and split the signal into two paths. One of the paths is then inverted, which results in two anti-phase square-wave signals being produced at the outputs of IC1c and IC1e. These anti-phase signals are fed to the inverting inputs of a dual operational power amplifier, IC2, via R3 and R4. The non-inverting inputs are connected to a potential divider chain which holds them at half the supply voltage.

Consider, now, the diode under test (DUT) being connected to the test terminals, TC1 and TC2, as shown. When IC2a output is high, current will flow through R5, D1, the DUT, LED2 and R6 to the output of IC2b, which will be at about ground potential. The result is that LED2 will illuminate.

During the next half of the cycle, the output of IC2a will be at about ground potential and that of IC2b will be high. Under these conditions, the DUT will be reverse-biased so no current will flow.

With the DUT connections reversed, it follows that the operation of the circuit

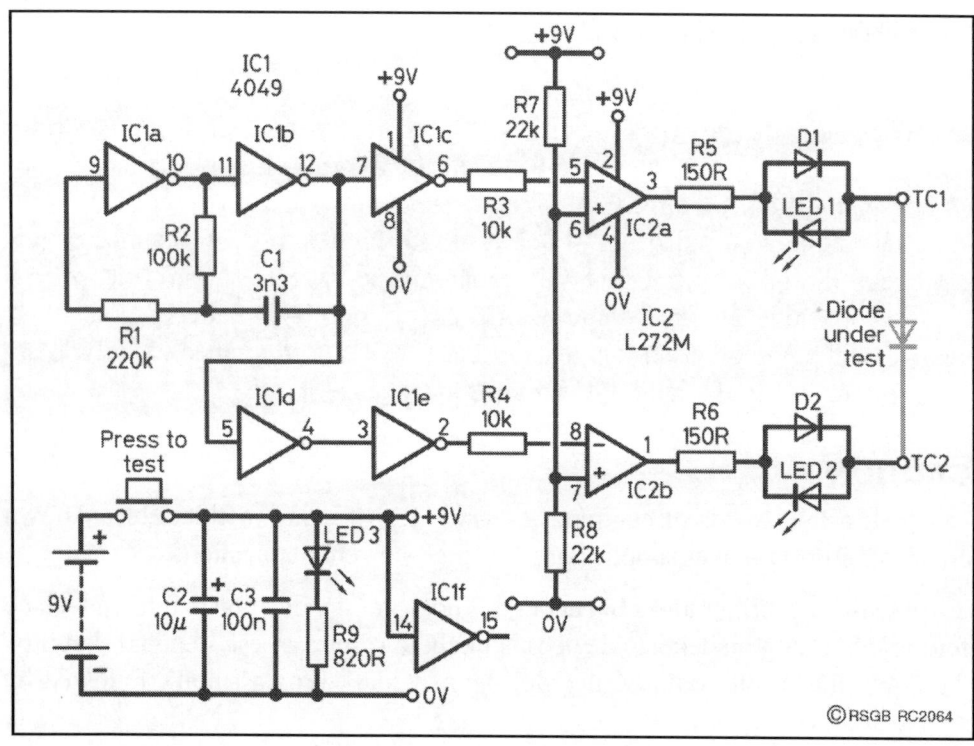

Fig 1. *The simple diode tester gives an instant 'dead' or 'alive' indication.*

will be similar to that previously described, but LED1 will now be illuminated instead of LED2. If the unit is constructed so that LED1 is adjacent to TP1 and LED2 to TP2, a direct indication is given of the terminal to which the cathode of the DUT is connected.

CONSTRUCTION

With the exception of the placement of C2 and C3, there is nothing at all fussy about the construction and the builder can use PCB, stripboard or 'dead bug' construction as preferred.

Whatever form of construction is used, C2 needs to be connected as close as possible to pins 2 and 4 of IC2, and C3 to pins 1 and 8 of IC1.

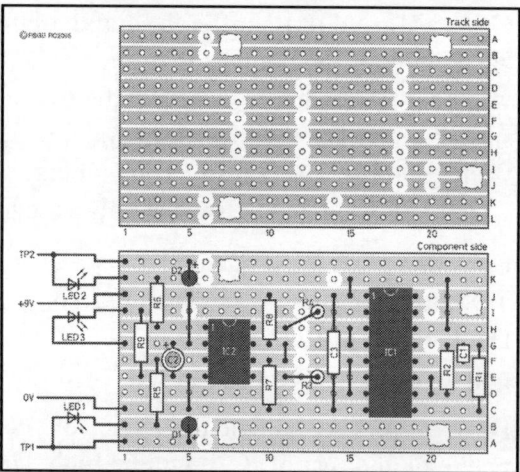

Fig 2. *Stripboard layout and component overlay. Note that if you use the recommended project box, there is little room to spare.*

A suitable stripboard layout is given in **Fig 2**. Some care is needed if the SB1A box is to be used. The width of the stripboard is such that it fits between the lid support pillars of the box with little room to spare. Check the placement of the board before drilling the mounting holes.

USING THE TESTER

Connect the diode under test between TP1 and TP2 and press the test button. You have a working device if either LED1 or LED2 illuminates, the illuminated LED indicating the cathode of the device. Should both LEDs illuminate at equal brilliance, the di-

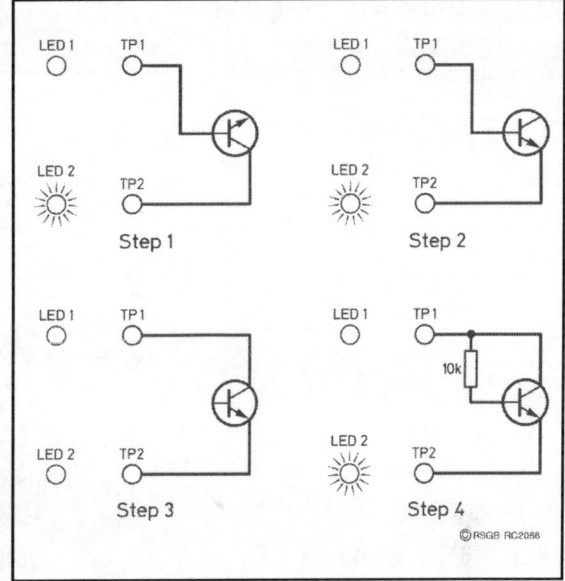

Fig 3. Use of the device to test bipolar transistors.

ode is short-circuit or is a low-voltage Zener. If both LEDs illuminate but one is brighter than the other, the chances are that the device being tested is a Zener with a breakdown voltage in the 4.7 – 6.8V region. In this case, the brighter of the two LEDs indicates the cathode.

If neither of the LEDs illuminates, this indicates either an open-circuit diode or a flat battery – check that LED3 illuminates to eliminate the latter possibility.

Testing transistors takes a few more steps – see **Fig 3**. Steps 1 and 2 can be skipped initially and carried out if needed to give information on the probable fault should step 4 fail.

At step 1 it might be found that both LED1 and LED2 illuminate, with one being brighter than the other. This is indicating that the reverse base-emitter breakdown voltage is less than about 7V, which is normal for some transistors. In this case the brighter of the two LEDs is the one to note. The transistor test is

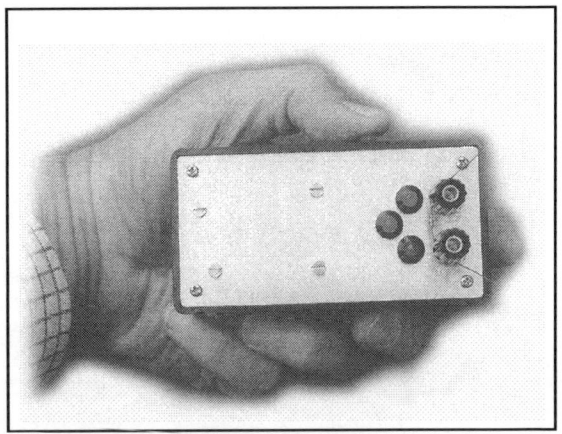

The completed diode / transistor tester.

not particularly exhaustive but it can be a simple way of checking that a device is still 'alive'.

PARTS LIST

Resistors - All resistors metal film 1/4W, 5%

R1	220k
R2	100k
R3, 4	10k
R5, 6	150
R7, 8	22k
R9	820

Capacitors

C1   3.3n ceramic (eg Maplin RA41U)

C2   10µ, 16V tantalum bead (eg Maplin WW68Y)

C3   100n polyester (eg Maplin BX76H)

Semiconductors

IC1	4049
IC2	L272M
D1, 2	1N4001
LED1	3 5mm high-brightness (eg Maplin WL84F)

Additional items

Push button switch, eg Maplin FH59P

4mm terminal posts, eg Maplin FD69A

Project box, eg SB1A – Maplin BZ27E

Stripboard

A Low-Voltage Alarm For Battery

Simple, direct-conversion receivers are prone to mains hum, and battery power prevents it. Assuming you use a rechargeable battery to power such a receiver, it should be recharged when the voltage falls below a predetermined level. Voltage checks can be made with a meter, but this can often be overlooked and the battery then becomes too flat to operate the equipment. This low-voltage alarm sounds when the voltage drops below a certain level to remind you that the battery needs to be recharged.

OPERATION

The alarm, shown in **Fig 1**, is connected across the receiver when in use. It is basically an oscillator circuit designed around a unijunction transistor, TR1. The emitter of TR1 is biased by the resistor R1 and the Zener diode ZD1. The circuit will oscillate when the emitter has a sufficiently low voltage on it, and will cease oscillation if this voltage rises.

ZD1 must be chosen to determine the alarm voltage you require, and some figures for your guidance are given in **Table 1**. To obtain alarm voltages between the values given in the table, adopt the following procedure. Let us suppose you want an alarm voltage between 10.9 and 12.1V. Choose the Zener diode voltage that gives a

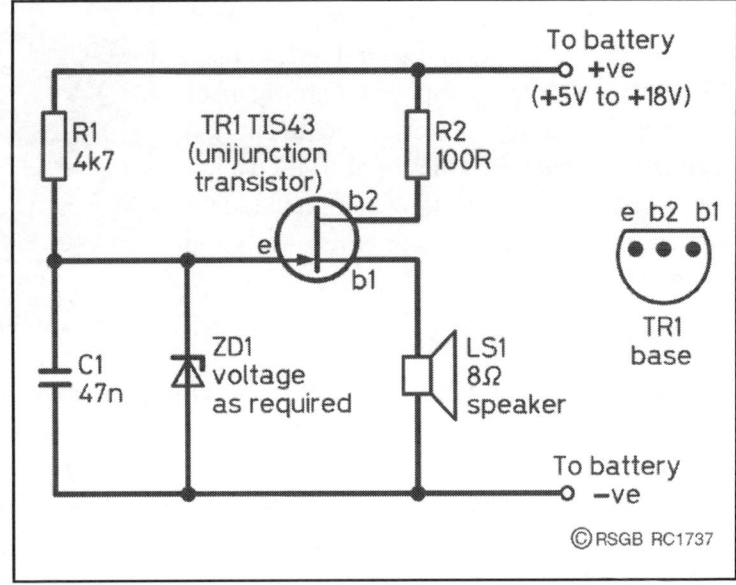

Fig 1. *The low-voltage alarm uses a unijunction transistor as an oscillator.*

10.9V alarm, and insert a forward-biased silicon diode in series with the Zener, the cathode of the silicon diode being connected to the cathode of the Zener. This effectively increases the Zener voltage to 7.2V (ie 6.6V plus the 0.6V across the silicon diode). This will raise the effective alarm voltage to between 10.9 and 12.1V.

Zener voltage Alarm voltage

Zener voltage	Alarm voltage
6.6	10.9
8.2	12.1
12	19

Table 1. A guide to the alarm voltages produced by various Zener diodes.

The alarm draws a small amount of current all the time so, although it could be left connected permanently across the battery itself, this is not recommended as it would eventually run it down.

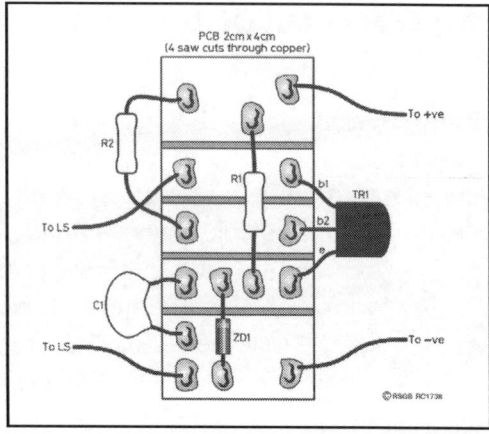

Fig 2. Physical layout. Before commencing assembly, make four saw-cuts through the copper of the PCB.

CONSTRUCTION

The circuit is built on a small piece of plain, single-sided PCB. Four saw-cuts are made through the copper, as shown in **Fig 2**, to make 'pads' to which the components are then soldered. When cutting, be careful not to cut right through the board, just the copper. I used a junior hacksaw to do this. The completed alarm can either be built into a case of its own or into an item of equipment.

The PCB can be fixed to the back of the loudspeaker with adhesive pads.

LIMITATION

The low voltage alarm relies on there being sufficient voltage in the battery you are monitoring to operate the alarm itself. Consequently, if the battery falls to such a low voltage that the oscillator will not operate, the alarm will not sound. Your receiver will not operate either, so you have two clues, both pointing to a flat battery!

PARTS LIST

Resistors	Capacitors	Additional items
R1 4.7k, 1/4W, 5%	C1 47n polyester	LS1 8-ohm, 1W
R2 100, 1/4W, 5%	**Semiconductors**	Case to suit (if required)
	ZD1 see text	PCB 2cm × 5cm
	TR1 TIS43	

The Old Rustic Paddle

Paddles for electronic keyers are usually a mechanical arrangement, often constructed with watchmaker precision. For portable or mobile operation such a paddle is easily damaged. Is there an alternative? Try this one . . .

THE RUSTIC PADDLE

This paddle has no springs or delicate mechanical parts. It comprises pieces of PCB material and flexible insulating material clamped between wooden blocks, and then fixed to a baseboard. The complete paddle is shown in **Fig 1**.

CONSTRUCTION

The paddle arm is made from two shaped pieces of PCB material stuck together-back-to-back; see **Fig 2**. The dot and dash contacts are made from pieces of PCB stuck to blocks of hardwood (65 × 30 × 15mm) using doublesided adhesive tape. Having done this, the three pieces are held together in their final position and two pilot holes are drilled to take the screws.

The holes in the three pieces of PCB and one in the wooden block are enlarged and elongated to clear the screws when they are inserted. Two pieces of dense sponge rubber (wetsuit neoprene is ideal!) are stuck as shown in **Fig 2** to the contact PCB. This makes construction easier, and the unit can then be assembled with the screws.

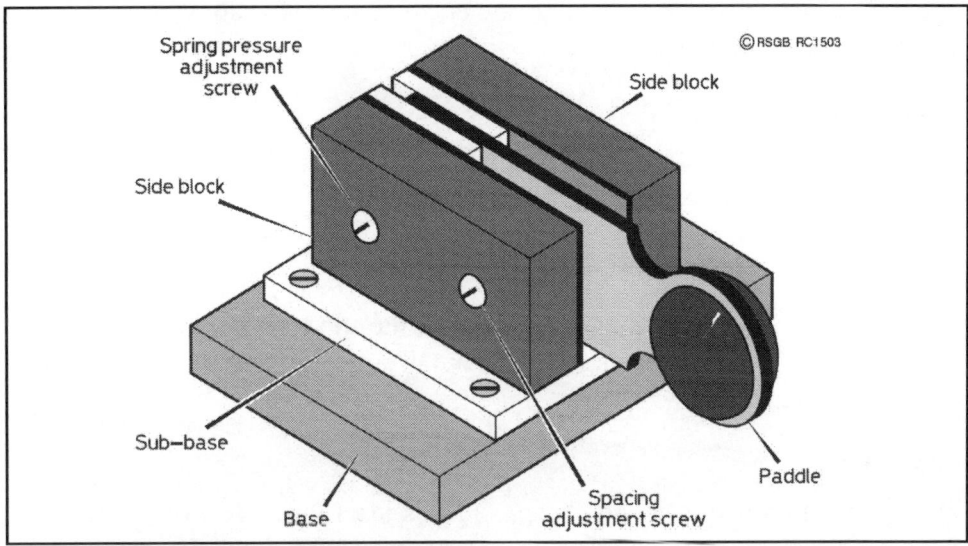

Fig 1. Construction of the complete paddle key

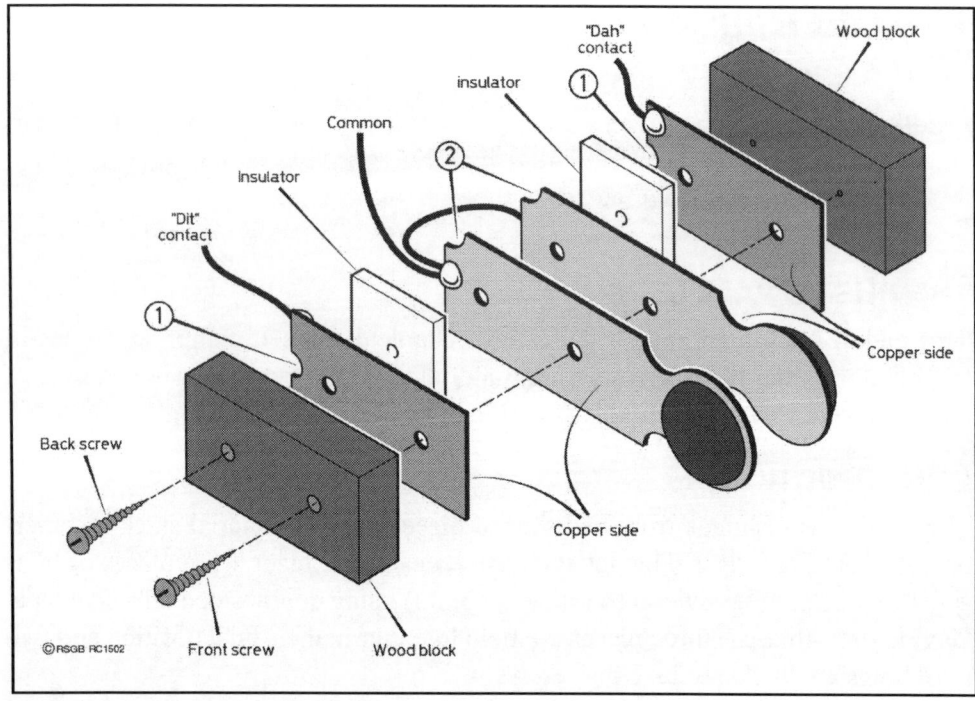

Fig 2:. *Exploded view of the paddle components. Cutouts (1) and (2) are to allow room for the solder blob connections on the facing PCB plates*

The wooden side blocks are fitted to a sub-base block; see **Fig 2**. Pilot holes are drilled in the sub-base block for the screws to hold the wooden side blocks.

The front holes in the sub-base block are enlarged; this is to allow movement for the paddle spacing adjustment.

ADJUSTMENT

Fix the paddles to the sub-base board but do not tighten fully. Connect the three wires to the electronic keyer. Adjust the 'spring pressure' using the rear screw, as shown in **Fig 1**, and the spacing using the front screw. These adjustments are interactive and may have to be repeated.

Tighten up the screws holding the paddles to the sub-base board and check the 'spring tension' and the spacing again. Fix the sub-base to a main baseboard as shown in **Fig 1**.

FINALLY

The construction is robust and cheap, and it is ideal for mobile or portable use. With a little more complexity (ie the two parts of the paddle operating independently) this idea could be converted for use with an iambic keyer.

A SIMPLE ELECTRONIC KEYER

A Danish amateur, OZ7BO, designed a clever and simple electronic keyer which became known as the El-Bug. This device used a double triode valve and a few small components. Some years later the design was updated by G3JIS using two transistors in place of the valve. Both designs used two relays, one of which is used to key the valve transmitter, which had a high positive voltage on the keying line.

These days, most transceivers have keying voltages of less than 12V positive to ground and the keying currents are low. It would be best to check your keying line prior to starting this keyer and, if it is a negative keying line or uses high current, a second relay could be fitted to the keyer output to key the transmitter.

The low power requirement of modern sets enables us to simplify the G3JIS version to 12 components, plus the bonus of a cheap and easily obtainable relay.

The El-Bug will not send automatic CQs or make the coffee but, handled properly, it will send perfect Morse at any reasonable speed after a little practice. It can be used with either a single- or double-lever paddle, but will not be iambic (ie it does not send alternating dots and dashes if both paddles are held together) and can be built on a 5cm × 6cm board without overcrowding.

CONSTRUCTION

The author used an assembly method which is a cross between the sawcut technique advocated by G3VTS ('TT', RadCom April 1995) and 'deadbug' construction favoured by G3ROO. All parts are mounted on the copper side of the board and the soldering pads are made large which simplifies construction. This allows for modification to be made with ease. To make a board like this is very easy indeed as no advance planning is necessary providing that more pads are made than you think will be needed.

Start by sticking PVC tape across the board in parallel lines, with 2 or 3mm gaps between them. Next, remove the unwanted tape with a sharp knife and the board is ready for etching in the normal way. No holes are required, as parts like relays and ICs can be stuck to the board 'dead-bug' fashion. The result will be as neat or ugly as you wish but I have built an SSB transceiver using eight boards constructed by this method and they have proved to be invaluable when modifications have been necessary.

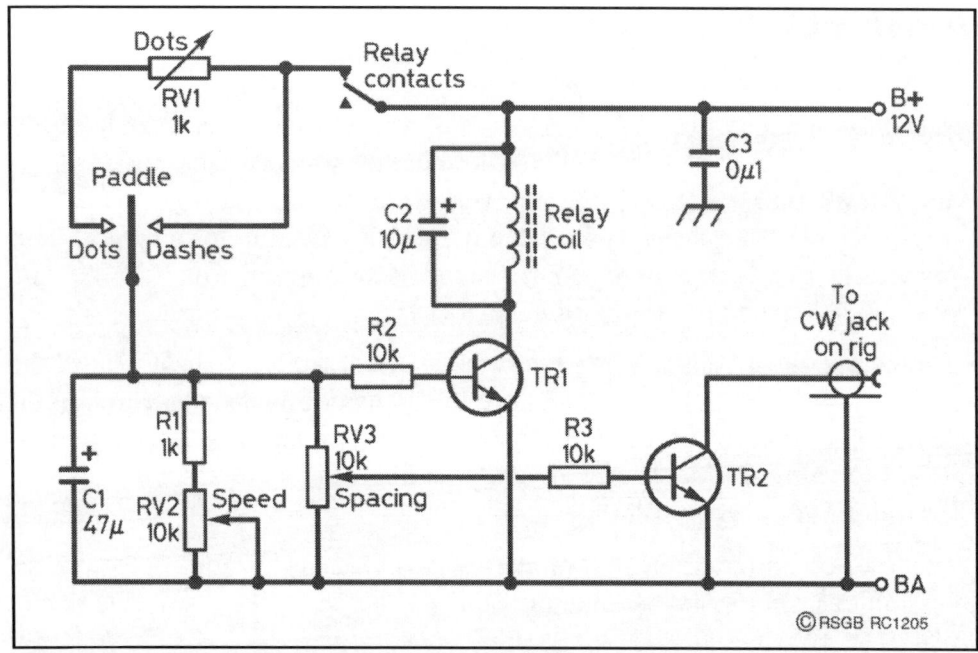

Fig 1: *Circuit diagram of the modified OZ7BO autokeyer.*

At first sight it seems hard to comprehend how such a simple circuit, shown in **Fig 1**, can generate dots and dashes of correct length and spacing without a binary digit in sight. To understand the operation, first note that the relay contacts used are in the normally-closed position. When the paddle makes contact with the dash side, the supply voltage is applied to C1 which charges, applying a positive voltage to both transistor bases. TR1 saturates, energising the relay and opening the contacts. At the same time, TR2 saturates and keys the transmitter. With the supply to C1 removed by the relay contacts, it discharges via the complex network connected across it. As the base voltage of TR2 drops below 0.6V, TR2 turns off and the transmitter is unkeyed.

At this moment, TR1 is still on due to its base voltage being slightly higher than that of TR2. As C1 continues its discharge, TR1 turns off and the relay closes again ready for the next operation. If the paddle is kept pressed a stream of dashes will be sent until the paddle is released. The difference in spacing is set by the time difference in the off states of TR1 and TR2. Due to the setting of RV3, TR2 is always switched off slightly earlier than TR1. A similar cycle takes place on the dot side, except that the quantity of charge in C1 is restricted by RV1, thus shortening the dashes to dots.

Correct operation of the El-Bug is very dependent on the characteristics of the relay. In fact it is the mechanical sluggishness of the relay that enables the circuit to operate at all. A 12V relay with a coil resistance of 400Ω is ideal. I use a miniature type YX94C from Maplin. This one is a little fast in operation, so I have retarded it with C2 connected across the coil. The value of C2 with the Maplin relay is not too critical, but a junk box relay may need some experimentation to get correct operation and a slower relay may not need any capacitance. The values of C1 and the three variable resistors are critical; do not leave out R1 or you may short out the power supply!

For initial testing, set the speed pot RV2 at half-track, which should correspond to about 12WPM. With an ohmmeter, set RV3 for 3kΩ between slider and ground. Connect a 12V supply, hold the paddle in the dash position and you will hear the relay clicking away quietly. Connect your rig into a dummy load and connect the key to the keying line. On pushing the dash paddle, dashes should be heard on the sidetone; adjust RV3 for optimum mark/space ratio. Now key the dot paddle and adjust RV1 for correct dot mark/space ratio. With the components specified, the keyer should work between 8 and 25WPM.

To prevent RF problems, the keyer should be built in a metal box and coaxial cable used to connect to the key jack of the transmitter. I had problems on 15m but this was cured by fitting two 10nF ceramic capacitors, one from the collector of TR2 to ground and the other between its base and ground. With these fitted, full-power operation was available on all bands.

OPERATING

A newcomer to auto-keyers may well find first efforts depressing, as the keyer seems to have a mind of its own. I remember 10 years ago borrowing a keyer for a few days and, although the marvel of it was obvious, I was unable to master it in five minutes and gave up. Now the beauty of making your own for a few pounds is the great psychological advantage you gain. Because you made it yourself and because it actually works, you feel obliged to persevere. You may well give up and put it away for a while but, sooner or later, you will go back for another go.

After three and a half hours' practice, I was able to do practice-QSOs at 12WPM and soon ventured on the air. I made mistakes, but slowly got better and the El-Bug has revitalised my interest in Morse and makes sending a pleasure. Paradoxically, sending good Morse has improved my receiving as well!

PARTS LIST
Resistors

R1	1k
R2, 3	10k

RV1	1k preset
RV2	10k linear potentiometer
RV3	10k preset

Capacitors

C1	47μ, 25V
C2	10μ, 25V
C3	0.1μ, 60V

Semiconductors

| TR1, 2 | BC108, BC109, BC171 or near equivalent |

Additional item

| Relay | Maplin YX94C |

Morse code tutor

A little something to make learning easier

Learning Morse code is never going to be easy and, unlike cycling, I believe it is something that you can forget. When I first learnt Morse code back in 1990 I had access to a Datong D70 Morse Tutor, which generated groups of five random Morse characters, either letters, numbers, or both. You could also control the speed of the generated Morse and the delay between characters. The portability of the device aided my learning as I could take it with me and make the most of a spare five minutes wherever I may have been.

For various reasons, amateur radio as a hobby got set to one side in my life for the best part of 23 years until late in 2013 when my enthusiasm for the hobby was re-kindled by being asked to take part in a Jamboree on the Air event. I spent the best part of 48 hours listening around the HF bands working stations across the world. When I returned home I knew that I had to get another radio but, more importantly, I knew that I had to be able to use the Morse key once again.

Sadly my old Datong D70 was no longer to be found in the house so I simply had to build one. It needed to be portable, have variable volume, variable tone, variable speed, variable spacing, and it needed to generate any combination of letters, numbers and punctuation. When I originally took the Morse test it wasn't a

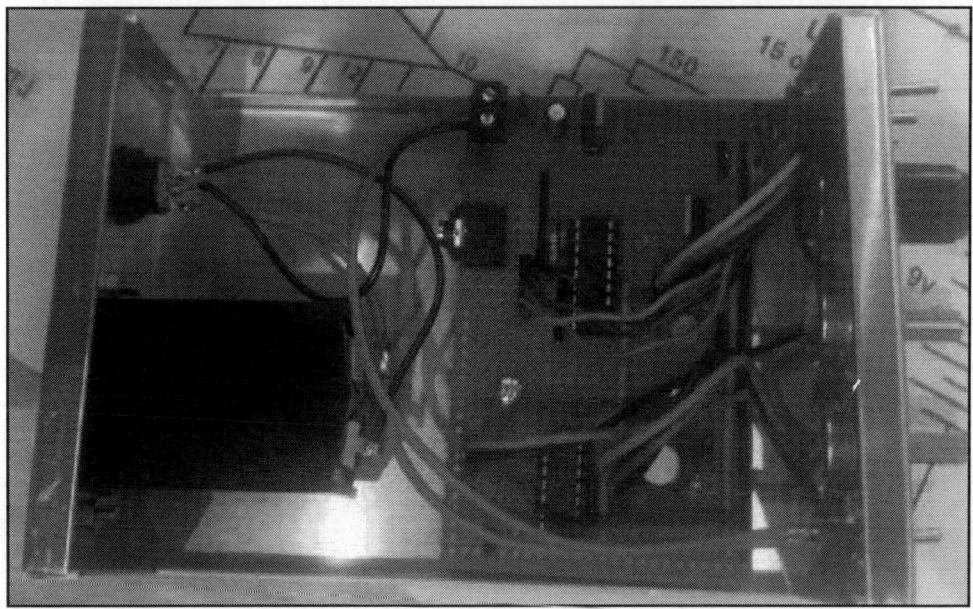

Photo 1: *Inside my prototype.*

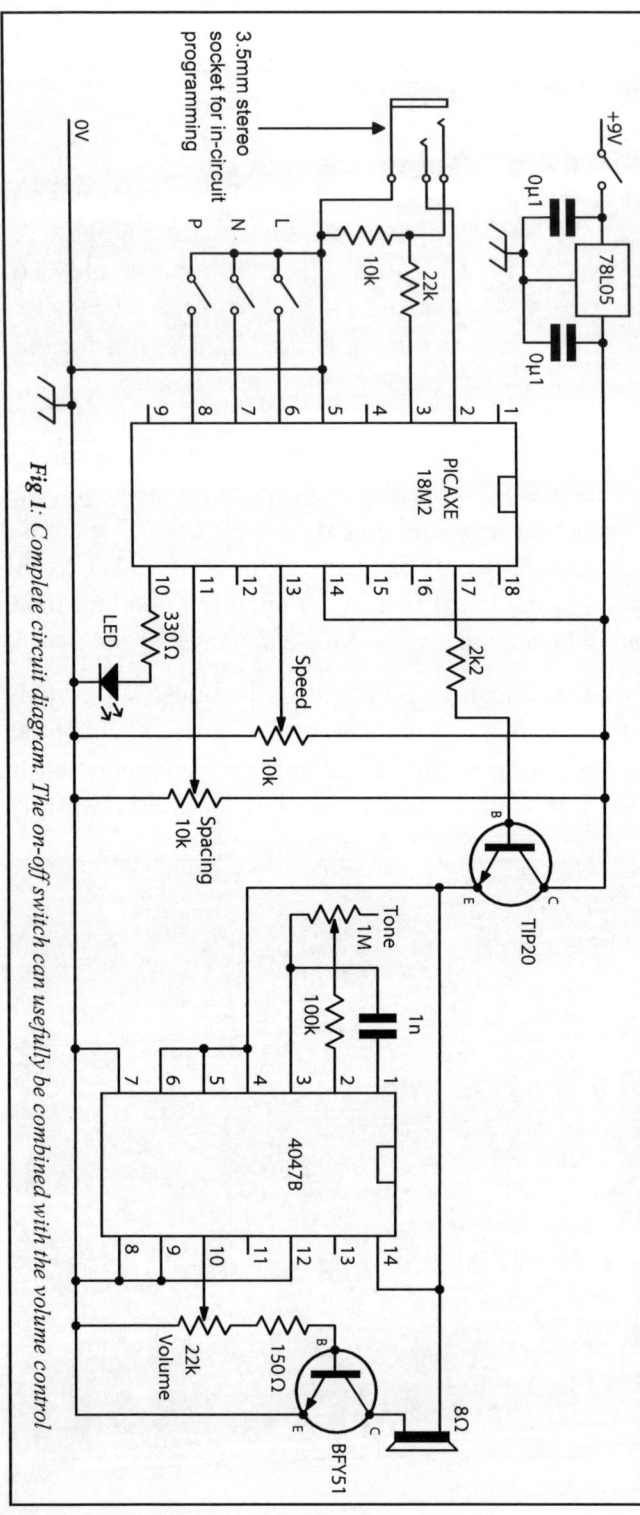

Fig 1: Complete circuit diagram. The on-off switch can usefully be combined with the volume control

requirement to learn any punctuation and I've always found it difficult to learn the extra characters.

HARDWARE DESIGN

The first choice was what processor I was going to use. One of the easiest embedded processors to use on the market today is the PICAXE. It's cheap, comes in various sizes with numerous input / output pins, it is programmed in a variant of BASIC – and I happened to have some in stock. I chose the PICAXE-18M2 as the starting block. Although the PICAXE chip supports tone generation I've never been happy with the quality of tone generated so I decided that the circuit would be in two distinct parts, the tone generator and the PICAXE controller, with the PICAXE simply applying power to the tone generator when required. Searching around the internet I stumbled over the Kent Morse Oscillator, based on the 4047B multivibrator. I built one to test; it sounded perfect and met my requirements of variable volume and variable tone.

The full circuit diagram is shown in **Figure 1**. To control the characters generated I made use of three simple switches: one each for letters, numbers, and punctuation. More than one switch can be on at anyone time to allow for different combinations of characters such as numbers and punctuation, or letters and punctuation, etc. Punctuation is interesting as there is punctuation in common use and punctuation in not so common use. I decided that if only the punctuation switch was on then it would generate the common punctuation and if all three switches were on it would generate all punctuation.

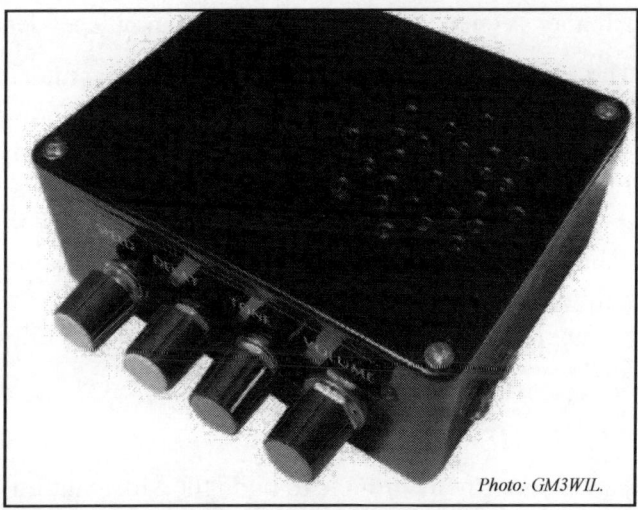

Photo: GM3WIL.

***Photo 2**: A second version, built to verify the design.*

Table 1: Supported Morse characters.	
All letters	A – Z
All digits	0 – 9
Common punctuation	+ - ,. / =? :
Not so common punctuation	" $ '( ) ; @ _ <Paragraph>!

Table 1 shows the supported characters.

To control speed and spacing I made use of the PICAXE inbuilt analogue to digital converters to read the voltage on the wipers of 10k potentiometers wired between 5V and 0V. This value was then used to index a lookup table. Using this method the speed can be controlled between 8 words per minute and 32 words per minute; the delay can be controlled between 0 seconds and 3 seconds – over and above the usual delay between characters that you would expect.

The circuit includes the necessary components to allow the PICAXE to be reprogrammed in circuit if necessary.

CONSTRUCTION

The circuit is very straightforward and there are no particular layout considerations. My prototype was built on stripboard and powered from a 9V battery (see **Photo 1**).

Another, seen in **Photo 2**, was built by an independent person to verify the design.

The most difficult part of building the Morse tutor will be programming the PI-CAXE chip itself. There are two alternatives: download the code from the author's website [1] and program the PICAXE yourself, for which you'll require PICAXE Editor 6 or similar on your PC (or AXEPad for the Mac), which can be downloaded free from [2] and a suitable USB or 9 pin serial to 3.5mm adapter cable, available from [3], to plug into the socket on the board. The second alternative it may be possible to contact the author directly, who possibly may be able to supply a ready-programmed PICAXE at cost price plus postage (whilst stocks last).

CONCLUSION

Having gone to the trouble to build the Morse tutor I'm sure that some of you will be asking, 'Has it worked?' Well, I'm pleased to say that it has certainly helped. I've been using it alongside local GB2CW transmissions by Chris, G3XVL, and I'm glad to say that I've recently got back on air and I am now making CW contacts once again. As for the future, you can always improve your Morse speed, so I'm looking forward to using the device for many years to come.

Martin D Waller, G0PJO

martin@the-wallers.net

WEBSEARCH

[1] www.g0pjo.the-wallers.net

[2] www.picaxe.com

[3] www.techsupplies.co.uk

A Simple USB CW Interface

When needing to send accurate CW for long periods of time, for example during contests, then it can be wise to do so by the use of a PC with appropriate software. Lots of software is available online, some of which is free, including Wintest, Logger 32, Winlog 32 or Mix W to name just a few. But how do you physically connect your radio to the PC for doing Morse code?

Well, a simple interface can be made that will allow your PC to send CW to your radio via a COM port using the DTR and GND connections of that COM port. This interface is easy to make and can be put togetger for under £5!

PARTS REQUIRED

What you will need to make the interface is the following:

Photo 1: A typical CP2012 UART converter.

- 1 x CP2012 UART converter. These are available from online sources such as eBay, Amazon, Aliepxress etc, often for under £1. **Photo 1** shows an example.

- 1 x USB enclosure – again, these are available online for under £1.

- 1 x 1.8m shielded cable, with 3.5mm or ¼" stereo plug already fitted, usually about £1.50 online. Buy whichever type suits your radio's KEY socket.

- You will probably also need a diode (1N4148 or equivalent) or a simple optoisolator, as discussed later.

- A soldering iron with solder and hot glue gun with some hot glue – but every amateur is likely to have these!

Photo 2: The RS-232 pin functions are clearly labelled. The red wire is connected to DTR (on the other side of the board).

CONSTRUCTION

The type of USB to UART converter you purchased and the size of the USB enclosure will decide if any work will need done. In order for me to fit my UART converter into the enclosure I used a rotary multi tool to take small shavings off the inside of the USB enclosure and also a small shaving off the UART converter itself.

The first step is to strip and tin the cable with your 3.5mm or ¼" stereo plug attached, ready for soldering. The connection from the tip of your stereo plug will be soldered onto the DTR pin of the USB / UART converter and the sleeve connection of the stereo plug

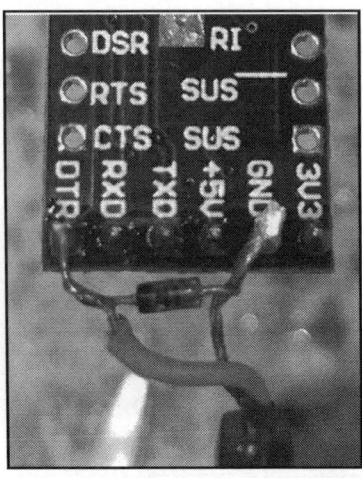

Photo 3: A diode fitted between DTR and GND, as per Figure 1.

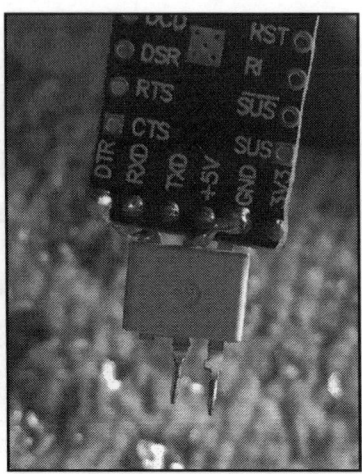

Photo 4: Fitting an optoisolator to the UART converter board.

will be soldered to the GND pin of the USB / UART converter, as may be seen in **Photo 2**.

Essentially these are all the connections that are needed, however in order to protect the devices a diode or 4N33 optocoupler can be fitted. *See* **Figure 1** and **Figure 2** for details. These show a DB9 connector instead of the USB interface I've suggested: as long as you get the pin identities correct (GND and DTR) you'll be fine.

The diode will fit easily within the USB housing, as can be seen in **Photo 3**. I just used a diode that I had lying around but anything similar to a 1N4148 would do.

Photo 5: Hot melt glue keeps everything nice and tight.

The optocoupler is relatively easy to graft onto the end of the converter, as shown in **Photo 4**. However, the resulting assembly is not so easy to fit inside the USB housing.

FINISHING OFF

Put the USB/UART converter into its housing and use some hot glue to hold the wires and connections in place, as shown in **Photo 5**. Once the glue has solidified, simply glue the two half of the USB enclosures together. **Photo 6** shows the finished article.

Once you have made up your interface, plug it into the PC and check the COM

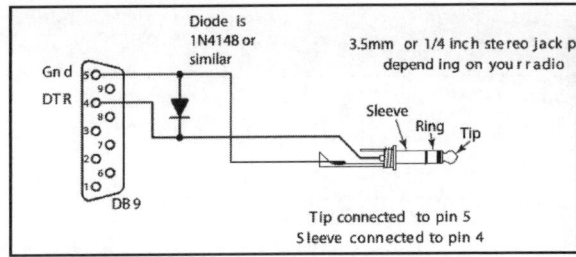

Fig 1: Basic connections between an RS-232 port (or USB UART converter) and radio.

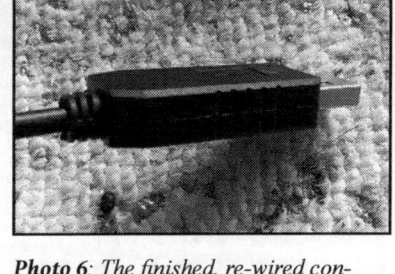

Photo 6: The finished, re-wired converter.

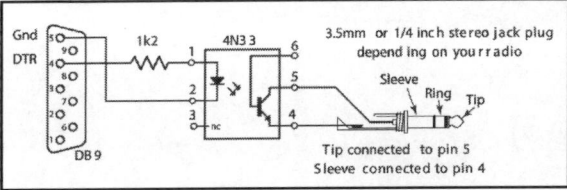

Fig 2: Better arrangement using an optoisolator to ensure electrical separation between the computer and radio.

port number. **Figure 3** shows how this should look in Device Manager, although the appearance will be slightly different in the various versions of Windows.

Once you have found out which COM port number your CW interface is allocated to, set up your software to show DTR as CW and, if RTS must be selected to satisfy your software, set that to PTT (although that's not relevant for this device). **Figure 4** and **Figure 5** show the key settings (in this case when my interface was on COM 7 – yours may well differ).

SETTING TO USE

Make sure you plug the stereo plug into your straight key socket on your rig if it has sockets both types (ie for paddle as well as straight key); this way you can still use your paddle and PC software depending what needs sending at the time. If you only have one

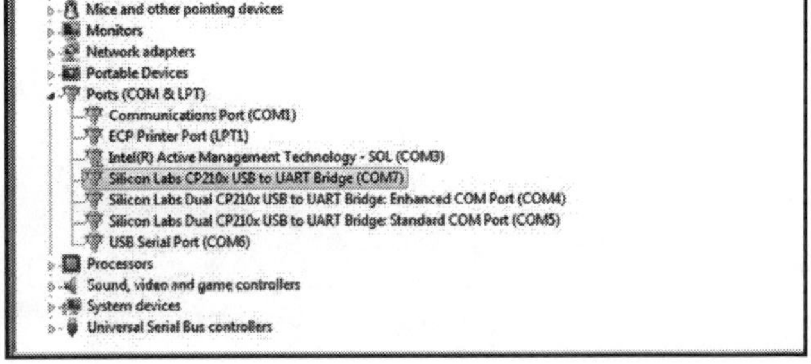

Fig 3: Typical way an USB to UART adapter shows up in a Windows Device Manager list (the exact appearance depends on the version of Windows you're using but the principle is similar).

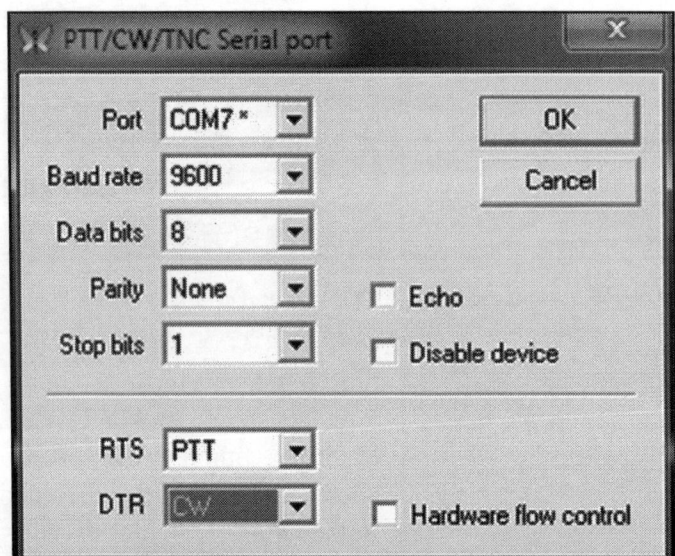

FIGURE 4: Typical settings for a PTT/CW/TNC serial port. The important bits are the COM port number, which must match your adapter, and the DTR = CW setting.

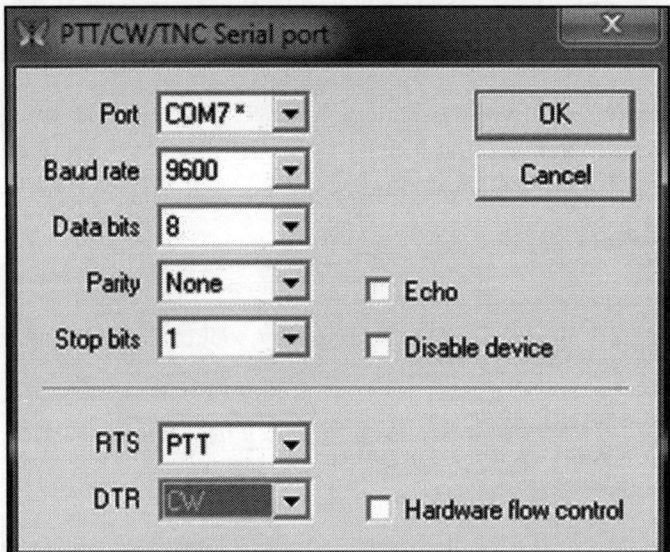

Fig 5: *Typical configuration screen for Morse software configuration. As with Figure 4 it's the COM port number and DTR = CW settings that are important.*

CW key socket on your rig, make sure the settings on the rig are correct for a straight key.

Billy McFarland, GM0OBX

gm0obx@yahoo.co.uk

A Simple Halo Antenna for the 2m band

The halo is a simple, easily constructed, lightweight and compact antenna that is horizontally polarised. Many amateurs start, and continue, on 2m with FM and a 'white stick' vertical antenna. These days many HF sets also have all modes on 2m, so why not give SSB (and CW) a try with a horizontally polarised antenna?

This is a very low-cost design that has given good results. It is ideal for loft mounting, or, with some weather-proofing, mounting outside. As this antenna is virtually omni-directional, there is no need for a rotator. This could be a good winter's evening project!

Photo 1: The completed halo antenna. Waterproofing would be required for permanent outdoor use

A small bit of theory

The halo transmitting element is basically a half wavelength, although it looks a lot smaller in the circular 'halo' format shown in **Photo** 3.1.

A standard half wave dipole has a feed impedance of around 50Ω at the centre. If you form this dipole into a circle, the feed impedance drops. However, tapping into a point away from the centre will find a point that is about 50Ω again. A gamma match is used to connect at this point, which is what the 'arc' next to the element is. A single, unbalanced, gamma match is fine – the whole element

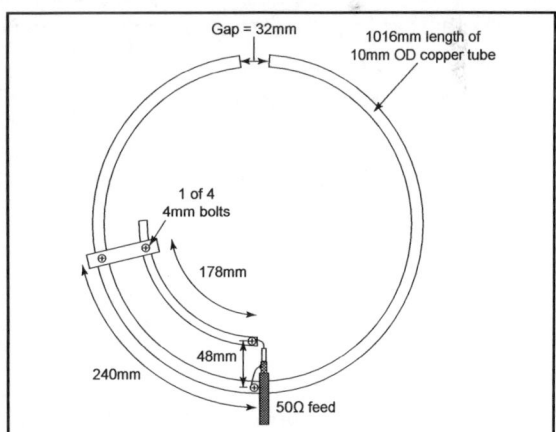

Fig 1: The component parts and general assembly arrangement

is still fed with RF energy. Be aware that, as with all half-wave dipoles, there is a voltage at the two ends. The halo has its ends near to each other at the gap – avoid touching the antenna at this point (or any other) whilst transmitting, or you may get an RF burn!

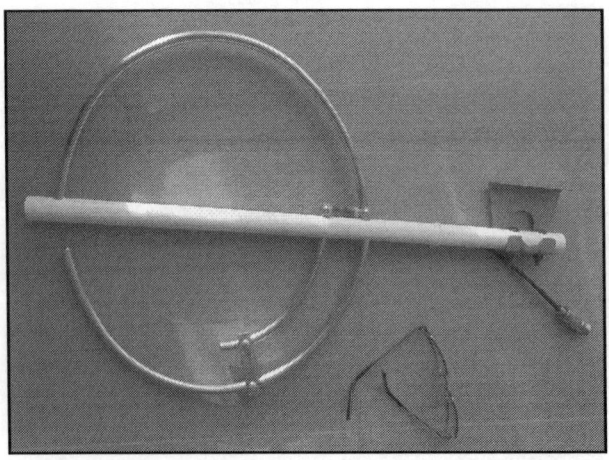

Photo 2: General arrangement of the halo antenna, with spectacles for scale

Some experimental notes

There are really three main variables in a halo design: the circumference, the end gap and the gamma feed point. Various sources in books and on the web were consulted and although the halo is a well-established antenna, there is quite a variance in dimensions. Therefore these three key dimensions were all varied at the 'lash up' stage and the ones given here were settled on.

Simple VHF dipoles and halos made of tubing, rather than wire, all tend to have a reasonable bandwidth. Although the circumference is key to the central operating frequency, once this dimension is determined and set there is no more to consider. The final circumference chosen here works well. Varying the gap can vary the resonance a bit, and varying the gamma point will also vary the impedance match a bit. Neither are too critical in this design and the measurements given have worked well.

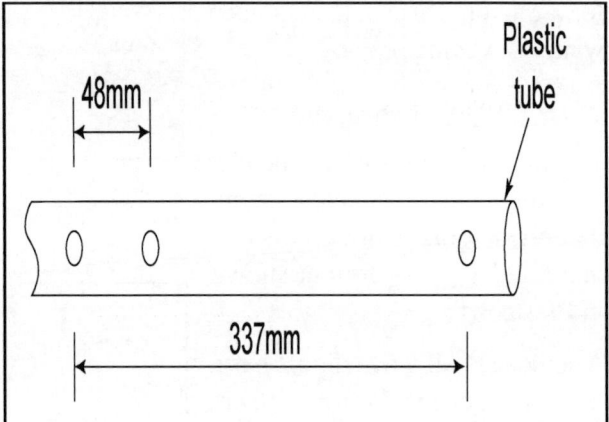

Fig 2: Drilling details of the support tube. Holes should be fractionally larger than 10mm

In some designs the gamma match is above or below the main halo element. Here it is in the same plane, which makes for simpler construction, although it can distort the theoretical radiation pattern.

Materials

The main material is copper tubing. This is not very critical, but here 10mm outside diameter microbore central heating tube has been used. It can be bought from DIY 'sheds' for around £25 for a 10m coil. This is more than you need for one 2m halo, but it leaves plenty for trying 4 and 6m designs – or perhaps a tuned loop for HF [1]. May be two people, or perhaps a group of club members, could share a coil?

10mm diameter copper tube of this sort is quite easy to form by hand and, as supplied, the coil is around the right diameter anyway.

Other items needed are:

- a plastic or fibreglass tube (I used a length of left-over 21.5mm diameter overflow pipe)
- some coaxial cable
- a connector (eg PL259)
- a bracket

some nuts, bolts and washers of about M4 size (suitable types can often be found in pound shops, more cheaply than at DIY sheds plus insulating tape and/or heatshrink sleeving.

Construction

Refer to **Photo 2** and **Figure 1** for details. Start by cutting a length of copper tube to exactly 1016mm. A small hand operated rotary pipe cutting tool is ideal and leaves a nice finish, but a hacksaw is also fine. De-burr the ends with a de-burring tool or a file and/or wire wool. Form this length into the best circle you can manage, which will be around 330mm diameter. Take time to do this. If there is a circular 'former', or some other radius item handy, then this can be useful. Don't worry too much – it only needs to be basically circular. Note that you must leave a gap of about 32mm between the ends.

Drill a hole at the centre point of the element, and another one 240mm along from it. The size of the holes should match the bolts you're using.

Next, cut a 205mm length of copper tube and form this into an arc to match the circle as best you can.

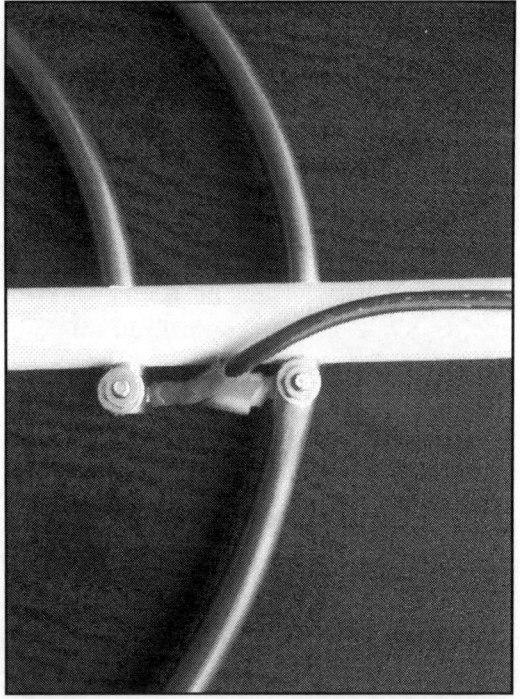

Photo 3: Close-up detail of the feed point arrangement

This will form the gamma match arc. Drill a hole close to one end, then another 178mm from the first hole, as shown. Note that this means 178mm around the tube, as shown by the dimension mark in **Figure 1**. Now cut a piece of copper tube 65mm long. Use either a club hammer or a strong vice to flatten the tube. It will form the shorting link. Drill 2 holes, as shown, 48mm apart. Ensure that all the metalwork has the rough edges removed.

The length of the plastic support pipe is not really critical, but around 600mm is probably about right. Drill three holes in it, as shown in **Figure 2**. These will need to be slightly greater than 10mm in order to give some clearance to the copper tube (but not too much). Probably the easiest – and safest – method is to drill 10mm holes and then open them out a bit with hand tools such as a round file or reamer.

Note that the end holes need only go through one side of the material. Finally, drill mounting holes at the end of the plastic tube, to suit your own bracket or mounting method.

Now fit the main element into the tube. Again, take the time to do this carefully. When satisfied, fit the gamma arc and the shorting bar in place. Adjust the halo gap again to about 32mm, if necessary. Refer to **Photo 1**. Finally, connect the co-axial cable, as shown in **Photo 3**, noting that the inner goes to the gamma match and the braid to the longer tube. Attach the coax to the antenna with crimped and soldered eyelets. It's probably easiest to use a short length of thin coax and terminate one end of this in a suitable plug or socket, then connect the main run of coax to this connector. For 2m, a coax run up to about 30' can be done in one of the modern RG8 Mini coaxial cables, without too much loss. Some are lower loss than others – check out the specifications.

Testing and installation

Either inside or outside, position the antenna in a reasonably clear spot and measure its SWR on 2m. You can do this on low power with a transmitter, or with an antenna analyser if you have access to one. The SWR across 144–146 MHz should be less than 1.5:1, with a figure near 1.0:1 mid-band. The SWR in the 146/7MHz experimental (NoV) segment should be below 2.0:1. Note that the SWRs quoted here are what you should be able to get when the antenna is in a fairly clear, high spot. A test in less favourable conditions may give a higher SWR. It is worth noting that an SWR of 2:1 only results in 11% of the transmitted power being reflected back to and this is generally OK for any transceiver.

If all is well, then mount the antenna in its final position. This halo is light enough to hang on strings in a loft, if you want to – see **Photo 4**. Outdoor mounting in the clear will give a substantial improvement for all VHF antennas. For outdoor mounting you will need to waterproof it a bit. Some lacquer over the screw connections would be sensible and some self-amalgamating tape over the main connections is also needed. Make sure that there is no chance of water getting into the coaxial cable. Also, if an in-line connection is used (as described here) then use self-amalgamating tape over the connectors.

Photo 4: The halo describe here makes a great more-or-less omnidirectional horizontally polarised antenna

Try listening to one of the 2m beacons [2]. Compare it against an existing vertical antenna if possible. Listen around in the SSB segment – roughly 144.300MHz ±100kHz. An ideal time for a good test is one of the RSGB's Tuesday activity nights – see the Sport Radio column for details.

Performance and conclusion

Simple tests showed that although it was not truly omnidirectional, in the sense that the signal is the same at all points of the compass, the antenna did have a fairly decent response all round. The power rating, assuming that the coaxial connections to the halo are sound, should be fine up to 50W (and probably beyond this).

This design ought to be scalable, so, for example, for a 6m design the key measurements could be scaled up by a factor of 145/51, that is multiply everything (except tube diameter) by 2.84. Note that if you build a 6m version and want it in the loft, then you'll likely need to do the final assembly in the loft, as the finished item probably won't go through the hatch!

References

[1] for example, 'A minimal loop for QRPp', RadCom, January 2016

[2] listed in the RSGB Yearbook or online at www.ukrepeater.net

A J-Pole Aerial For 50MHz

For FM communication (ie voice and data) on the VHF bands, a vertical aerial is used to give all-round (non-directional) coverage. This is a half-wave aerial which can be fed at the end, thus removing the principal problem with the conventional vertical centre-fed half-wave dipole, which is that the feeder should leave the dipole at right angles. This is no problem when the dipole is horizontal, but can be difficult for the vertical dipole.

BASIC FACTS

A feeder must be connected to an aerial at a point where the impedance (AC 're-sistance') of the aerial closely matches that of the feeder. The difference between the two impedances gives rise to the voltage standing-wave ratio (VSWR), which is unity only when the two impedances are the same. With 50W feeders, the feed-point of a half-wave aerial is at the centre, where the aerial impedance is around the same value. At the end of a half-wave aerial, the impedance is high, so it is not a suitable point to connect a 50W feeder.

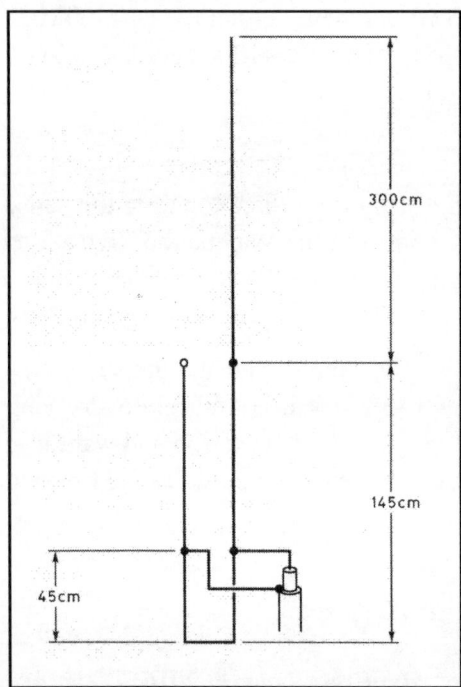

300cm

145cm

45cm

Fig 1: Overall dimensions of the 6m J-pole aerial.

Connection at this point can be effected using an RF transformer. RF transformers act in the same way as ordinary transformers, except that they are much smaller, and usually comprise wires of particular lengths adjacent to each other. **Fig 1** is a good starting point. It shows the aerial in its diagrammatic form. Notice that the aerial is in the form of an elongated letter 'J'; this shape gives rise to its nickname – the J-pole. The quarter-wave RF transformer is the lower 'U'-section below the half-wave element. At the bottom of the U-section, the impedance is zero (this may become clearer later) and at the top of the U-section it is high, thus matching the aerial impedance. The co-axial feeder cable is connected part-way up the U section, where the impedance is around 50W.

THE PRACTICALITIES

Fig 2 shows the aerial as constructed. As it is about 4.5m high, it may be too high for the average house loft, but is ideal for mounting outside, supported by a non-metallic pole or hung from a tree branch. The upper half-wave section is made from 1.5mm insulated copper wire, as used in domestic mains wiring. The quarter-wave transformer be-low the half-wave section is made from 300W balanced line ('ribbon cable'). The wires at the bottom of the transformer section are stripped of their insulation, twisted and soldered. At the upper end, only one wire of the balanced line is soldered to the bottom of the half-wave section. The other wire of the pair is not connected and is left insulated.

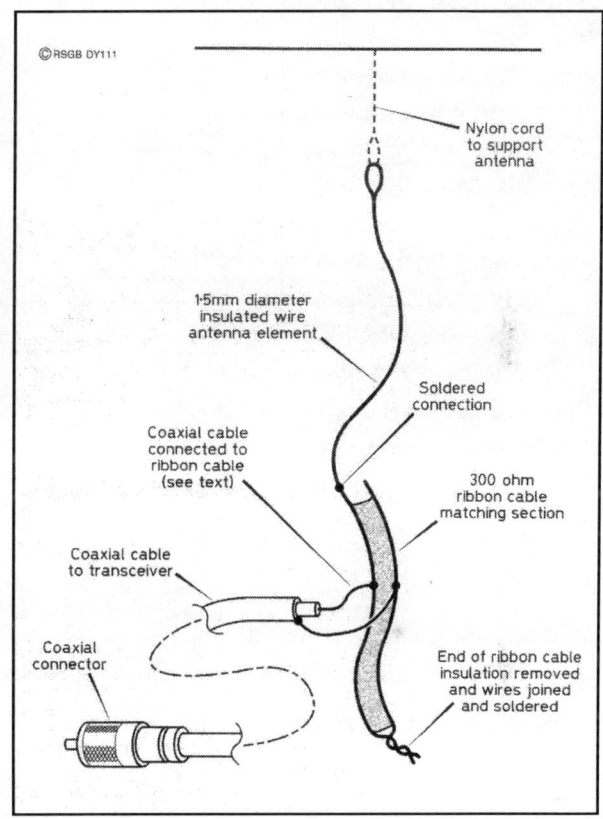

At the feed-point of the trans-former, the insulation needs to be carefully stripped from the balanced line. You will need a standing-wave meter (VSWR meter) in the coaxial line between your transmitter and the aerial, and you will need to adjust the position of the feed-point; 45cm from the bottom was the best point on

Fig 2. Constructional details.

the prototype, but this position is dependent upon the immediate surroundings of the aerial, and must be done when the aerial is in its final operating position.

Warning: Never make adjustments to the feed-point when the transmitter is on. Make a VSWR measurement, switch off, move the feed-point, switch on again, make another measurement, and so on. You will need to aim for the lowest VSWR; you can certainly get better than 2:1. Having found the best position, wrap all the exposed wires with self-amalgamating tape, to seal them against the ingress of moisture.

HOW IT PERFORMS

Fig 3 shows a computer prediction of how the J-pole radiates. It is called a polar diagram, and shows the distribution of your transmitted power when viewed 'from the end of your garden'. Most of your signal is sent at a fairly small angle to the horizontal; very little signal goes upwards, which is a good thing, of course. This also shows why the

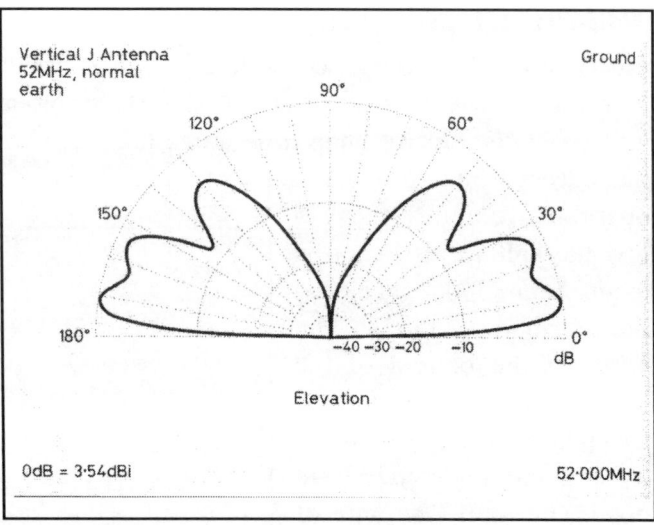

Fig 3: The computed polar diagram of the J-pole.

J-pole (or any other vertical aerial) should not be called 'omnidirectional', which means it radiates in all directions. It is omnidirectional only in the horizontal plane.

SAFETY

Where you mount your aerial is a matter of personal preference and the restrictions of height and space, but the following safety rules must be applied.

1. Never fix an aerial where it may come into contact with power lines or telephone lines.

2. When climbing a ladder to put up an aerial outside, make sure the ladder is safe and that it is secured.

3. Don't do this alone. Preferably have someone with you. If this is not possible, make sure someone knows where you are.

PARTS LIST	
3.00m	1.5mm insulated copper wire
1.50m	300W balanced line
As required	50W coaxial cable
As required	Self-amalgamating tape

A Portable Three-Element 6m Yagi Antenna

So, you've just acquired a transceiver with HF and 6m coverage, and you are keen to try out the 'magic band' for the first time. What better way than with a simple, portable antenna, which you can dismantle and use for some /P operation during the summer months?

REQUIREMENTS

When contemplating the addition of another antenna, most people would list their requirements as follows:

- Boom length, maximum 2m
- Reasonable gain
- Low cost
- Ability to dismantle easily for transportation
- Easy assembly with the minimum of tools
- Easy and quick tune-up
- Lightweight

First, materials are needed. A visit to the local DIY shop should provide a good source with a wide range of 1m and 2m lengths of aluminium tubing and box sections. Use 12.5mm diameter tube in 2m lengths for the centre sections of the elements, and 10mm for the element end sections. These sections of tube fit snugly. You will also need a 1m length of 10mm rod to make the gamma match section. A 2m length of 25.4mm square U-section aluminium was used for the boom.

THE DESIGN

A three-element beam always has one reflector, one driven element and one director. For sim-

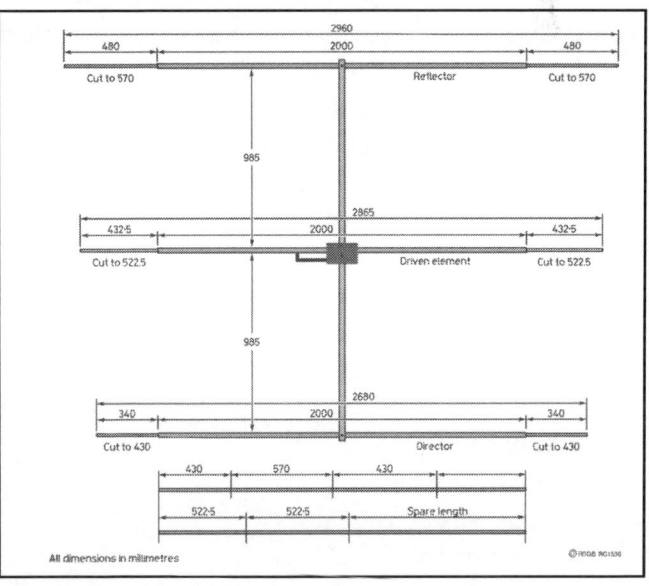

Fig 1: Dimensions and tube-cutting directions.

plicity, 'plumbers delight' construction is employed so all the elements and the boom are at a common earth potential. This reduces some of the static electricity which can be prevalent in other types of design. However, this limits the feeding arrangements to a delta or a gamma match. The gamma match was chosen because of its easy adjustment, bearing in mind the requirements for quick tune-up and transportability.

DRILLING THE BOOM

The first thing to do is to measure out the boom and drill the holes for the reflector and director elements (see **Fig 1** for dimensions). The 12.5mm holes are marked

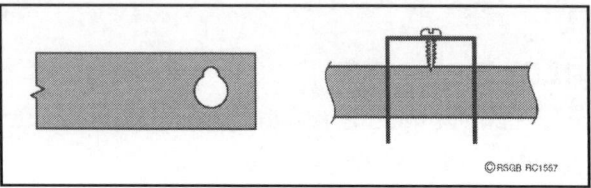

Fig 2: Details of the element holes (reflector and director).

and drilled with a smaller (6mm) drill bit before the larger holes are drilled. Care must be taken to get these holes straight and level in both directions, otherwise the antenna will look very odd indeed. Using a small round file, file a notch in the top side of each of the four holes. The notch should be just large enough to pass the head of the screw which holds the 10mm tubing to the 12.5mm tubing (see **Fig 2**). The reason for doing this is simple; it makes it easier to dismantle or assemble the elements on the boom. In the flat side of the boom drill a 2mm hole on the element centre line; this is for the screws to hold the elements in place (see **Fig 2**).

DIRECTOR AND REFLECTOR

The 'standard' sizes of tubing available are perfect for a 6m beam. The author bought two 2m lengths of 10mm tube and carefully cut them to the dimensions shown in **Fig 1**. In this way one piece gave me the correct lengths for four end-pieces (reflector and director). With these end-pieces cut to the right lengths, they were simply inserted into the 2m lengths of 12.5mm tubing and, after marking the correct length of the whole element, 2mm holes were drilled for the self-tapping screws to be driven through both the 10mm and 12.5mm tubing to hold them together. Each screw was driven through the tubing in a similar manner on opposite sides. This is done on the director as well as the reflector (see **Fig 3**).

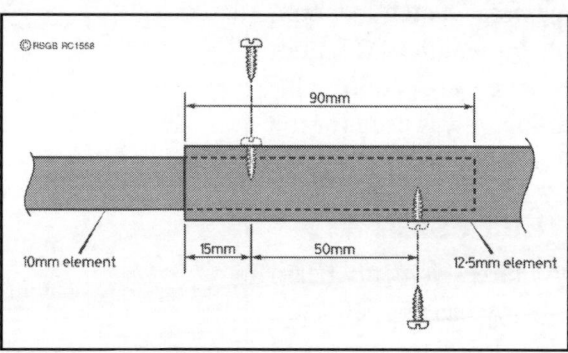

Fig 3: Detail of joining end-pieces to the main elements.

DRIVEN ELEMENT

The driven element is made of the same materials as the reflector and director. It is fixed to the top of the boom together with a plastic weatherproof box, which houses the 50pF capacitor and the gamma arm assembly.

The fixing screw through the box into the boom allows the driven element 12.5mm tubing to go in parallel to the boom for easy transportation.

At this stage, the 10mm tubing and the 12.5mm tubing are not yet fixed together. This is so that the 12.5 mm tubing will fit tightly into the connections box. Holes must be drilled in the 12.5 tubing of the driven element for the gamma arm. These holes allow connection of the gamma arm by a screw inside the tubing. A larger hole (8mm) is drilled through one wall of the tubing, and a 2mm hole is drilled in the opposite wall. The reason for the 8mm hole is to allow a screwdriver access to tighten the element-to-gamma arm screw. After fitting and testing, the 8mm hole can be filled with putty or simply taped over with weatherproof tape. To get the correct location of the gamma arm holes, first find the exact centre of the driven element and then measure out 305mm – this is the centre-line for the holes. At the centre of the element, another 2mm hole must be drilled for the shield connection of the coaxial cable.

CONNECTIONS BOX

The connections box needs to be drilled to take the driven element, gamma matching arm and the capacitor shaft. The first holes to mark and drill are for the driven element. These were carefully marked on each end of the bottom half of the plastic box, then drilled with a 6mm pilot drill and then again with the 12.5mm bit. The gamma arm hole is drilled 40mm away from the element hole, and is only drilled in one end of the box. The driven element and 1m aluminium rod are placed into the box and a suitable position for the capacitor located and marked for drilling. The capacitor came from the junk box, and the exact position is not critical as long as it does not foul the element or gamma arm (see **Fig 4**). A hole is required to screw the connections box to the boom; this is a 2mm hole drilled in the centre of the box. This screw goes through

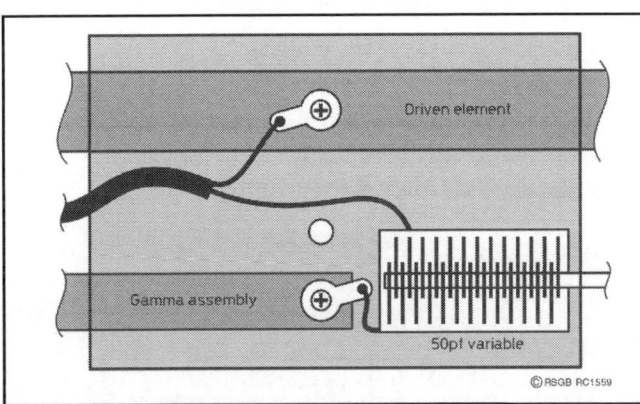

Fig 4: *Connections from the coax to the gamma match.*

the plastic box and into the boom from the top side. The other hole required at this time is one for the coaxial cable. This is drilled in the same end as the gamma assembly hole in the centre of the box.

GAMMA ARM

With the 1m length of aluminium rod, measure and cut 320mm off. At one end, drill a 2mm hole for the connection point of the capacitor. 90° out from this, at the other end of the rod, drill a 2mm hole straight through the rod. Using a 6mm drill bit, drill half way through the rod to

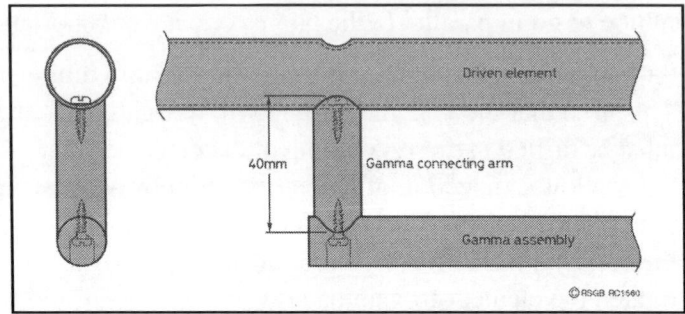

Fig 5: The gamma arm assembly.

make a countersunk hole in the rod (see **Fig 5**). This is the attachment point for the gamma connecting arm.

From the left-over length of aluminium rod cut a piece exactly 40mm long. With a file, one end must be filed to fit the gamma arm, and the other end filed to fit the 12.5mm tubing.

If a drill-press is available, it can be used to make a perfect job by drilling through the rod with the correct size of drill bit. However, the file also works well, though it does take a bit longer. In each end of the gamma connecting arm a 2mm hole is

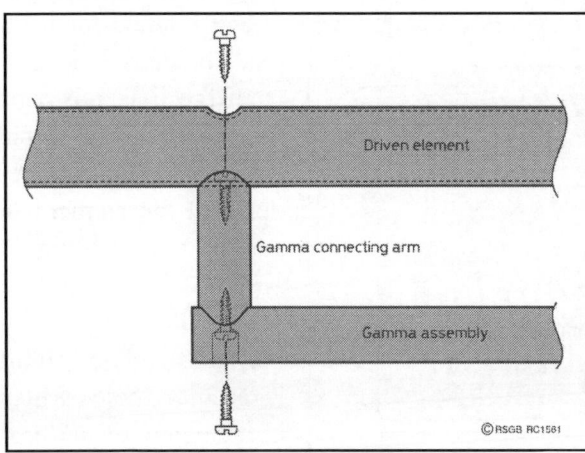

required for the fixing screws (see **Fig 5**).

Using a self-tapping screw, the gamma connecting arm to the gamma arm, forming an 'L'-shaped gamma assembly. Place a screw through the 8mm hole in the driven element and screw the gamma assembly to the element using a self-tapping screw (**Fig 6**).

Fig 6: Connecting the gamma arm assembly to the element.

Slide the connections box over the driven element with the hole for the gamma assembly facing the gamma assembly. Insert the gamma assembly into the hole in the box and adjust the box until it is in the centre of the element. Now screw the end-pieces into the driven element in the same manner used for the director and reflector (see Figs 4 and 5). Fit the capacitor to the box and connect one side of the variable capacitor to the gamma assembly, using the 2mm hole (drilled earlier) and a solder tag held in place with a self-tapping screw. Feed the coaxial cable into the box from the outside and strip the ends ready for connection. The shield of the coaxial cable is soldered to a solder tag and screwed to the driven element using a self-tapping screw. The inner of the coaxial cable is soldered directly to the other side of the variable capacitor. All the external connections to the box can now be weatherproofed using hot-melt glue, epoxy or similar product.

FIXING THE DRIVEN ELEMENT

Locate the centre of the boom and drill a 2mm hole in the exact centre of the flat edge (see **Fig 7**). The connections box is screwed to this point using a self-tapping screw. Now, turn over the boom and align the driven element so that the boom and the element are at 90° to each other. Drill through the boom and the plastic box and into the driven element with a 2mm drill bit. This hole will be off-centre and it holds the driven element in the correct place when the beam is being used.

WEATHERPROOFING

Fill the ends of all the elements with hot-melt glue, epoxy, or a similar long-lasting product to keep rain out of the tubing. After checking all the connections are good in the connections box, close the lid and seal against the weather. If required, drill holes in the side of the boom for a mast clamp.

CHECKING THE SWR

The VSWR should be quite good if the dimensions in this article are followed closely. Place the assembled antenna in a clear area at least 3m off the ground and check the VSWR on a known VSWR bridge. The VSWR can be adjusted by turning the capacitor and checking again. It is best to

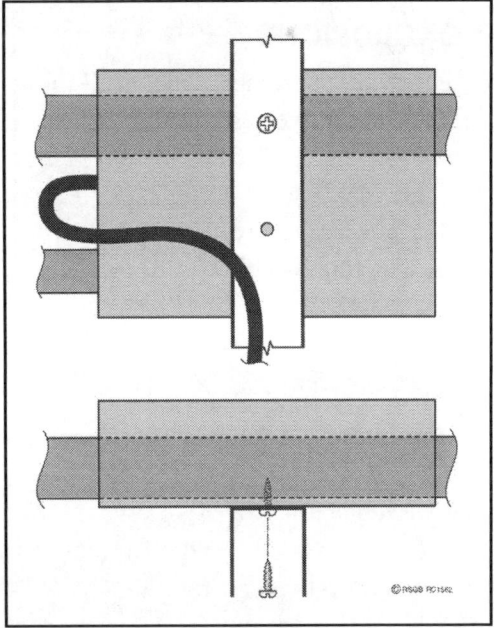

Fig 7: Connecting the box to the boom.

check the VSWR at both band edges and set the VSWR minimum at the centre of the band (see **Fig 8** for VSWR readings on the three beams built for testing purposes).

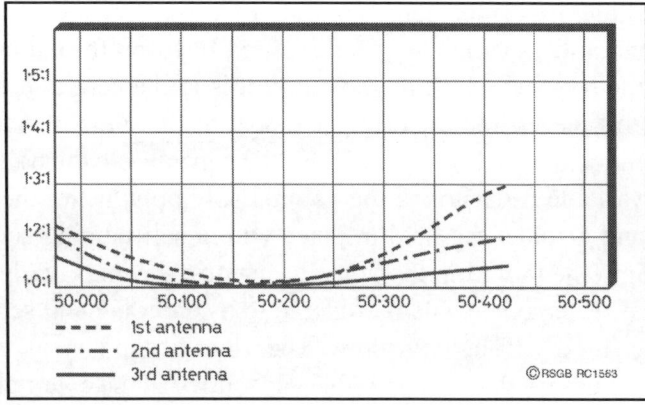

Fig 8. *VSWR readings on the three beams built for testing purposes.*

ASSEMBLING AND DISMANTLING

Lay out all the metal pieces in a clear flat area. Rotate the driven element to 90° and insert the retaining screw through the boom and into the driven element. Slide the director through the boom and lock it in place using a screw through the boom and into the element. Do the same thing with the reflector. The complete antenna is shown in **Fig 9**.

To dismantle the beam, lower the mast, remove the antenna and the three screws in the boom. Remove the director and reflector and rotate the driven element through 90°. Only a single screwdriver is required for the beam and three screws.

CONCLUSIONS

The three-element beam meets all the criteria which were set out at the beginning. It has proved robust enough for everyday use at home as well as portable use on my 10m pump-up mast. The total weight for the antenna is less than 3kg.

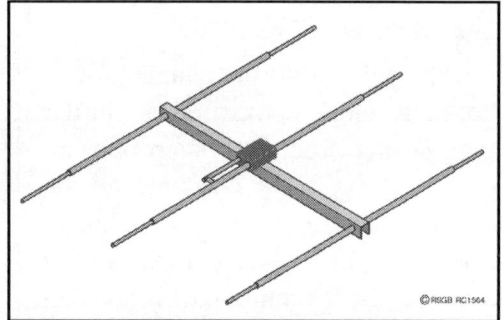

Fig 9. *The complete three-element 50MHz beam.*

PARTS LIST

Boom section	One 2m length, 25.4 × 25.5mm square 'U' aluminium
Element centres	Three 2m lengths 12.5mm OD aluminium tubing
Element ends	Two 2m lengths 10mm OD aluminium tubing
Gamma match	One 1m length 10mm aluminium rod
Connections box	Plastic box 70 × 122 × 50mm
VC1	Variable capacitor, 50pF
Other items	Solder tags, self-tapping screws

Two-Element, 6m Quad Aerial

6m – the 'magic band' – where you can work across town or try your hand at some real DX during the summer months. This simple aerial is robust – more so than a three-element beam in an exposed, windy situation, and it performs well.

CONSTRUCTION

This is simple and uses a chunk of 5mm aluminium plate, nylon cable ties and glass-fibre rods. **Fig 1** details the construction.

Having drilled the metal plate and loosely fixed the rods in place with cable ties, holes are drilled about 6mm from the ends of the rods and 90° from the plane of the loop. By doing this, the wire will not slip when under tension, keeping the symmetry of the loop. The wire lengths are carefully measured, adding 30mm for soldering the ends. The loops are made up from normal multi-strand hook-up wire, although thicker wire could be used without materially affecting the performance. The driven loop and reflector have wire lengths of 543cm and 594cm respectively.

A gamma match is made up on a piece of Perspex using the earth wire out of 2.5mm mains cable for the loop. The wire is folded into a 'hairpin' 31cm long and with 2cm spacing (**Fig 2**). The thickness of the wire and its small size enables it to be supported only by a piece of Perspex at the capacitor and feed-point. About 20pF is required for the capacitor, so a 35 or 50pF variable is used to allow for variations.

The reflector loop is fed through the holes in the ends of the four longer rods and the two ends are soldered together. Next, the driven wire is attached to one side of the gamma match, the free ends passed through the holes of the shorter rods and the loop completed by soldering. In the final quad, the gamma match capacitor is in a small plastic box to shelter it from the rain.

Finally, the rods are slid in or out of the cable ties so the 700mm loop spacing is correct and then the cable ties are pulled tight, locking the structure in a stable, if not rigid, configuration. It can be made more rigid

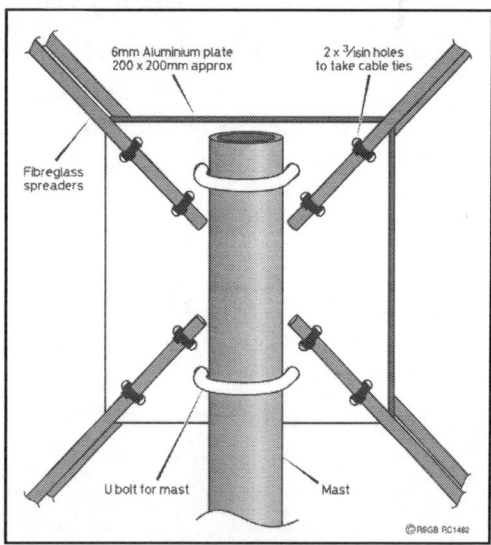

Fig 1. Glassfibre spreaders are fixed to an aluminium plate with cable ties.

by tying the ends of the driven loop to the corresponding reflector loop using fishing line.

SETTING UP

Various feed systems were used, the best using 72Ω twin feeder, but the complexity of the balanced ATU in the shack left a lot to be desired, so the second best feed using a gamma match is described.

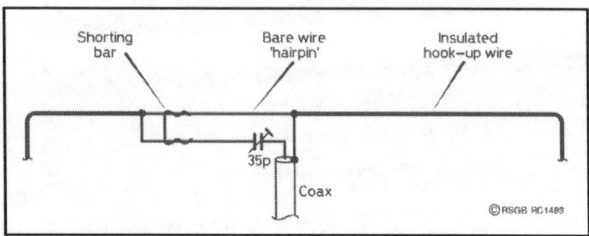

Fig 2. Details of the gamma matching system.

The gamma match consists of a coupling wire and capacitor. The tapping point and capacitor are adjusted to give a perfect match to the feeder, measured on a VSWR meter. A match can be maintained provided the feeder is kept perpendicular to the horizontal part of the driven element and equidistant to the vertical sections. During testing and setting up, you will see the VSWR rise rapidly if the feeder is moved from one side to the other, demonstrating the importance of keeping aerial systems symmetrical! The gamma match may be adjusted by standing on a [wooden] step ladder with the base of the quad about 8ft above ground. A shorting bar of wire is moved 5mm at a time along the hairpin, starting from the closed end, and the capacitor re-trimmed for best VSWR. This is repeated until the point of best VSWR is obtained. When mounted on the side of the house, the resonant frequency rose by 100kHz and the VSWR rose to 1.2:1 due to the change in location.

The quad aerial is easy to construct if the right materials are at hand and it is easy to maintain. The glass-fibre rods used were spares, but they are available from garden centres for making cloches. Failing this, bamboo canes could be used, but these must be selected for thickness and flexibility so that the final shape is satisfactory.

PARTS LIST
Glass-fibre rods, 8 or 10mm diameter, 4 × 108cm long and 4 × 111cm long
Aluminium plate, 20 × 30 × 0.5cm
2 × 50mm U-bolts. Car exhaust clamps are ideal
12 × 5mm nylon cable ties. Make sure you have spares
Variable capacitor, 50pF
Copper wire, 1m solid, 2.5mm
Multi-strand hook-up wire, 12m

Warning! Never make any adjustments to an aerial when the transmitter is running. Always switch off the transmitter, make your adjustments, and switch on again – Ed.

A 26 element Yagi-Uda beam for 23cm

> The purpose of this project is to demonstrate that it is possible for a newcomer to the hobby to build useful equipment for any band from LF to microwaves. Access to advanced test equipment may be useful, but it is not always a prerequisite for success.

The goal will be to build a high-gain 23cm beam using materials that are available from the local DIY store.

There are many different types and configurations for VHF/UHF beams. There are good reasons why the Yagi-Uda array is such a popular choice for amateur constructors and commercial manufacturers. The 'Yagi' is a very simple, almost two-dimensional structure made from an array of dipole elements. Compared to more elaborate 3D structures like parabolic dishes or beams based on full-wave loops (Quad or loop-Yagi), the standard type of Yagi is relatively easy to design and build. The simplest configuration consists only of a series of perfectly straight conductors made from wire, or more commonly at VHF/UHF, metal rod or tubing. This type of structure is particularly well suited to computer modelling. Problems with designing or modelling/simulating a Yagi are usually related to the mounting hardware for the elements and/or the feed system. For example: if the elements are mounted on a conductive metal boom, it may be necessary to make corrections to the element length. The amount of correction required will depend on whether the element is mounted directly on, close-to or through the centre of the boom. To complicate matters even further, a different correction factor must be applied depending on whether the elements are connected directly to the boom or electrically isolated using plastic insulators.

The authors preference is to build VHF/UHF beams on a non-conductive wooden or plastic boom. The aerial is first simulated on a computer. Once satisfied with the performance of the computer model, the hardware version is built and tested. To minimise the risk of error, the design should be as simple as possible. I try to avoid the use of unnecessarily complex structures like fancy feed,

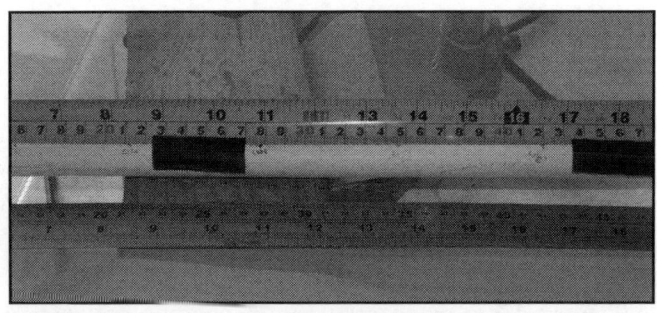

Photo 1: Measuring the boom for drilling.

matching or phasing networks, or anything that is likely to stretch the capabilities of the modelling software. VHF/UHF/microwave simulations are usually done for 'free space' rather than near the ground or any other conductive structure. I have found most of the available software to be very reliable for accurately modelling the behaviour of very simple wire structures like dipoles or quad loops.

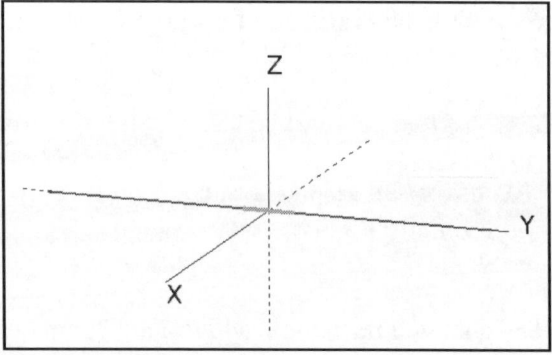

Fig 1: 1m of wire from Y -0.5 to Y +0.5.

NEC-2 [1] and similar software describes wires by defining the start and end points as X, Y, Z coordinates. For example, a 1m long horizontal wire could be described as: 0 -0.5 0 0 0.5 0

There are no specific rules regarding wire orientation. The usual convention is that Z is height, X and Y are the horizontal planes. **Figure 1** shows our 1m wire as shown in *Xnecview* by PA3FWM [2].

Once you become familiar with the way *NEC* defines wires, you can easily design more complex structures made from mutiple wires. A more complete definition of our 1m wire is GW 1 11 0.0 -0.5 0.0 0.0 0.5 0.0 0.0005

This tells NEC that this is wire number 1, to be analysed as 11 separate segments, with XYZ coordinates as previously described and a wire diameter of 1mm (radius = 0.0005m). If this is to be used as a centre-fed ½ wave dipole, the feedpoint will be segment 6 at the centre point of the wire. A 1m free space wavelength is equal to a frequency of 300/1 = 300MHz. The half wave resonance for this dipole will be a few percent below 150MHz. The *NEC-2* plot shows resonance at 145MHz and very credible values for gain and feedpoint resistance of 2.13dBi and 71.5Ω respectively. This is a good example of how *NEC* can provide accurate and reliable results.

Design

The Yagi-Uda array (commonly known as the Yagi) was invented by Shintaro Uda in the 1920s. In most cases, when a transmitter is connected to a Yagi, only one element is directly driven from the feedline. The remaining elements in the array are parasitically coupled. The phase and magnitude of signals in the parasitic elements is determined by the spacing between elements and by the element length. Uda's clever design places the elements so that radiation tends to be re-inforced in one direction and cancelled in the opposite direction. As the Yagi is a passive device, there is no real power gain. The apparent gain is because most of the available power is focused in one direction. The last couple of Antennas columns have explained how these antennas work. In a well-designed Yagi, gain is proportional to the boom length or to the number of elements used. Considerable gain is possible at VHF and above where elements are relatively small and boom length of several wavelengths are easily achieved.

Back in the good old days, Yagi designs were refined and optimised by experimentation. This cut-and-try approach is extremely time consuming and requires access to a testing range where your design can be compared to a known reference. Since the 1970s, *NEC* (Numerical Electromagnetics Code) software has been used to model the performance of aerial designs using a computer. Several versions of *NEC* have been developed over the years. The most commonly available is *NEC-2* and its derivatives. *NEC* has been updated, modified and ported from the original Fortran for other languages like C and C++. I used *nec2c* [3] for the examples shown here.

NEC has led to a revolution in Yagi design. It is now quite easy to come up with a design that is optimised for your individual requirements. A typical Yagi design involves trade-offs between several important parameters like gain, bandwidth, front/back or front/side ratio, tolerance to construction errors and so on…

A few basic rules apply: gain is proportional to length, so you can't achieve high gain simply by placing a large number of elements on a short boom. Higher gain for a given boom length can be achieved by using more than one Yagi in your array. Vertically 'stacked' Yagis are more commonly used than horizontally separated 'bayed' arrays, presumably because it is convenient to use a single vertical support mast. Large arrays of stacked and bayed Yagis are often used for EME (moonbounce) operation. Additional gain will always be less than the theoretical maximum of 3dB for each doubling of the number of aerials because of losses on the phasing/feed networks.

Photo 2: Balun made from RG79 75Ω coaxial cable.

A conventional Yagi will usually have a single driven element, a single reflector and perhaps one or more directors. Early designs tended to use equal spacing between all elements and all directors cut to the same length. Modern 'long boom' designs are a little more complex, with unequal spacing between elements and gradually decreasing director lengths as more elements are added. The long boom Yagi designs by G Hoch, DL6WU were a significant milestone in Yagi design. Even now, several decades after they were first published, The DL6WU based designs are still very close to the state of the art.

A high gain Yagi for 23cm

This month's project is a 23cm Yagi. The elements and boom were made using only materials from the local B&Q. I took the lazy approach and used the K7MEM on-line calculator [4] to calculate the element lengths and spacings. The design frequency is 1297MHz. Options chosen were DL6WU spacings, non-conductive boom, gain ~16dBd. Boom length and element diameter were dictated by the available materials.

The boom is a standard 2m length of white 20mm plastic electrical conduit. Using DL6WU spacings, this length limits the number of elements to around 26. Each element is made from a short length (around 100mm) of 2.5mm copper wire. This was stripped from standard 2.5mm twin and earth mains cable. Note that 2.5mm is the cross-sectional area and not the diameter. The element lengths and spacings are shown in **Figure 2**.

Construction

The insulation was removed from the 2.5mm wire to leave bare copper

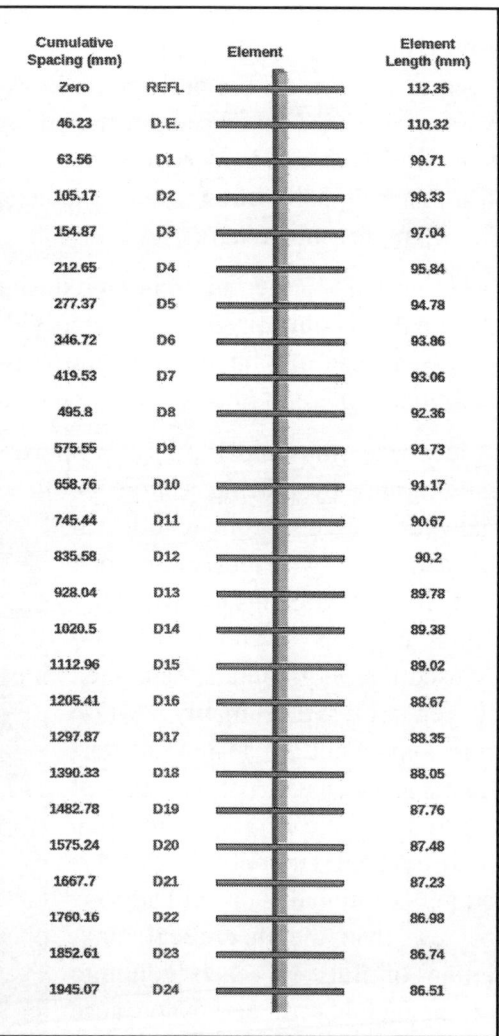

Cumulative Spacing (mm)	Element	Element Length (mm)
Zero	REFL	112.35
46.23	D.E.	110.32
63.56	D1	99.71
105.17	D2	98.33
154.87	D3	97.04
212.65	D4	95.84
277.37	D5	94.78
346.72	D6	93.86
419.53	D7	93.06
495.8	D8	92.36
575.55	D9	91.73
658.76	D10	91.17
745.44	D11	90.67
835.58	D12	90.2
928.04	D13	89.78
1020.5	D14	89.38
1112.96	D15	89.02
1205.41	D16	88.67
1297.87	D17	88.35
1390.33	D18	88.05
1482.78	D19	87.76
1575.24	D20	87.48
1667.7	D21	87.23
1760.16	D22	86.98
1852.61	D23	86.74
1945.07	D24	86.51

Fig 2: *Element lengths and spacings of the DL6WU design (see text).*

with a diameter of 1.78mm. Element lengths were cut about 0.5mm too long and then filed down to the exact length.

Be careful when handling the boom because of the risk of eye injury to anyone nearby.

Element positions were marked carefully using a fine marker. The measuring tape was secured to the

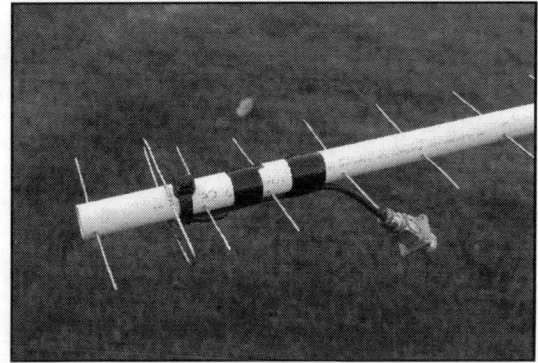

Photo 3: Elements at the rear of the boom.

boom using a few strips of insulating tape. I used a 1.5mm drill for the element holes. **See Photo 1**. As the final (front) few cm of the boom are surplus to requirements, I drilled a 5mm hole through the extreme end of the boom and pushed a length of 5mm aluminium rod through this hole. The rod was laid flat on the workbench to keep the boom stable while drilling the 1.5mm holes for the elements vertically through the boom using a drill stand.

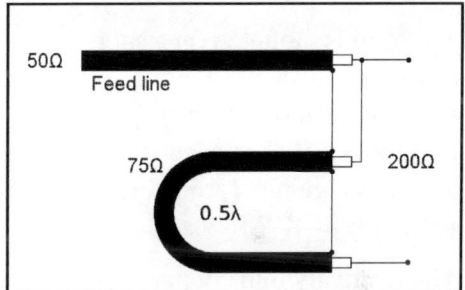

Fig 4: 50Ω to 200Ω balun made from a 0.5λ section of 75Ω co

The aerial is designed for a 50Ω feed when a centre-fed dipole is used as the driven element. I found it more convenient to use a folded dipole with ~200Ω impedance and a 4:1 balun made from a short length of coax. The driven element is also made from 1.78mm bare copper. Details are shown in **Figure 3**. The 4:1 balun is made from a half wavelength of 75Ω coaxial cable. I used RG59 (Maplin XS52G or similar). This cable shows a measured velocity factor exactly in agreement with the claimed specification of 0.66. The 0.5λ balun section is (150÷1,297)x0.66 = 76mm long. *See* **Figure 4** and **Photo 2** for details of the balun.

Photo 4: The completed 23cm Yagi.

I used a very short length of RG58 50Ω coax and an in-line N-socket for the connection to the main 50Ω feeder to the shack. As feedline losses can be quite significant at 23cm, it is advisable to use low-loss cable. I used a 12m length of Westflex 103.

The elements were pushed through the boom using pliers. As the holes are smaller than the element diameter, it took a fair bit of effort to push them through. As the elements are such a tight fit, I'm confident they will remain in place. If in doubt, or if you will be mounting the beam with vertical polarisation, you can secure the elements with a small amount of epoxy resin adhesive at each side of the boom. The driven element is placed around rather than through the boom. The folded dipole is slightly splayed at the centre to give a good fit on the boom. The balun section is held in place with a cable tie and the short length of RG58 is taped to the boom. Feedpoint impedance is critically dependent on the spacing between driven element and first director. This arrangement allows for fine-tuning of SWR by making slight adjustments to the position of the driven element. **Photo 3** shows the rear of the boom.

The Yagi is mounted on a short fibreglass stub mast using a couple of heavy duty cable ties and vinyl tape. The boom is stabilised from above and below using a length of nylon cord in a 'diamond' configuration. The assembled Yagi is shown in **Photo 4.** An *Xnecview* plot of the *NEC-2* radiation pattern is shown in **Figure 5**. Maximum gain is 18.6dBi or 16.45dB which is in close agreement with the value predicted by K7MEM.

The relatively thin copper elements won't withstand much mis-treatment (such as large birds using them as a perch), so it's probably best to treat this as a portable or test antenna.

Websearch

[1] www.nec2.org/
[2] *http://wwwhome.ewi.utwente.nl/~ptdeboer/ham/xnecview/*
[3] www.qsl.net/5b4az/
[4] www.k7mem.com/

VISTA Aerial

> The Variable Inductance Small Telescopic Antenna – nothing to do with computers!

WHY SMALL?

The fun of low power (QRP) communication has much to do with achieving more with less. The same goes for experiments with small antennas. If they can be carried about it is easy to accept a degree of ineffi-ciency in exchange for convenience of use.

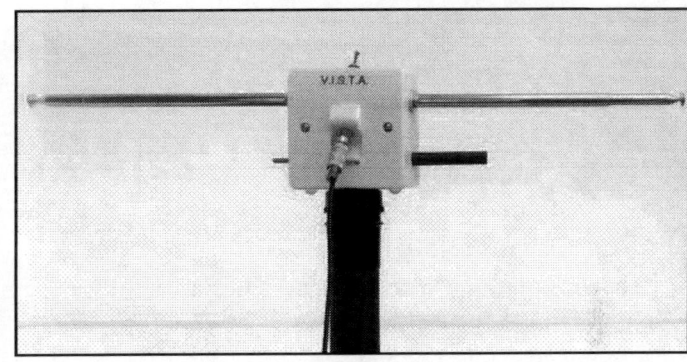

Photo 1: *The complete VISTA antenna, shown with the two 1.3m long 10 section telescopic dipole elements retracted. Extended, the antenna is just over 2.6m wide.*

This compact antenna design uses a ferrite rod and so it is suitable only as a low power experimental device. Whilst not an efficient radiator, it is very handy, needs no pole or guys to keep it up and requires no separate matching unit. It also needs no counterpoise or earth connection.

HOW IT WORKS

The VISTA is a centre loaded shortened dipole. The ferrite rod increases the 'Q' of the inductor but it is unclear whether the small associated increase in radiation resistance is of practical use. Some RF radiation is detectable with an absorption wavemeter along the coaxial feed to the VISTA, but the strongest field is around the inductor and proximal third of the telescopic elements. As the feed line falls away at right angles from the dipole there should be minimal interaction between them.

ORIGINS

The idea came from a simple antenna matcher [1] that used a ferrite rod to vary the inductance of a self-supporting coil. It occurred to me that something similar might be useful to resonate a small dipole. Initially I took two telescopic lecture pointers (each about 60cm long) and wound a loading coil at the centre. The coil was loop coupled via a length of RG174 coaxial cable to a bidirectional (Stockton [2]) power meter and a SoftRock/Power SDR transceiver tuned to the 14MHz band.

As the ferrite rod brought the system to resonance there was a gratifying increase in the basal noise trace and several CW signals could be tuned in on the receiver. Even more encouragingly, with a series capacitance of about 80pF included in a central coupling link, it was possible to achieve a 1:1 match to the 50Ω transceiver output. With minor changes, similar results were obtained on each band from 7 to 28MHz.

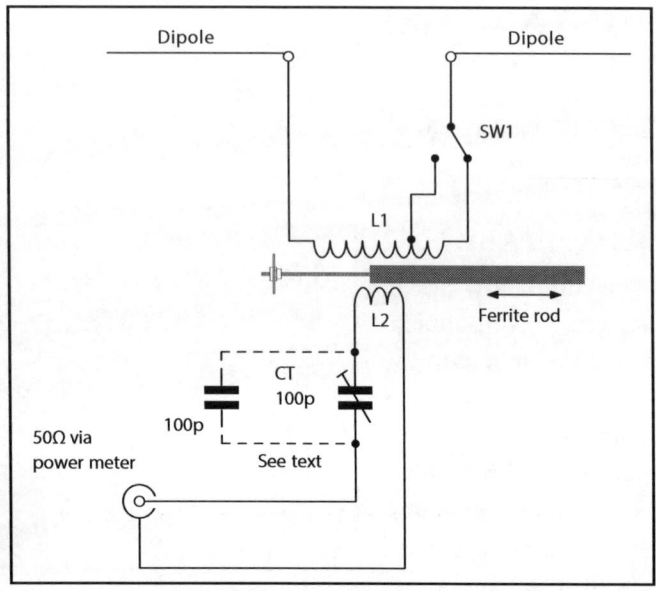

Fig 1: *Circuit diagram of the VISTA antenna. See text and Table 1 for component and coil winding details.*

Still with the short 'pointer' elements in place and with the antenna in a ground floor room, I heard a strong CQ call on 18MHz from a station near Kiev, Ukraine. No one answered, so I called him back. He copied my callsign and gave a 339 report. Output power was 2W and the antenna was just 1.3m above floor level.

It seemed worth exploring further, so I built two separate units for the HF range. More switched taps on the coil would have allowed wider coverage but I was keen to keep things as simple as possible. I later found that the series capacitor could be omitted if a single turn coupling coil was wound at the 'ferrite end' of the loading coil.

CIRCUIT

The circuit of the VISTA antenna is quite simple. The central components are L1 and L2 that are coupled to a varying degree by the ferrite rod. Variable capacitor CT provides matching to the incoming 50Ω line (supplemented by an additional 100pF for the lower frequency version). L1 directly drives the telescopic elements of the antenna, with a couple of taps at one end selected by switch S1. **Figure 1** shows the circuit diagram and **Table 1** gives component values for the low-and high-band versions.

The main difference between the low-and high-frequency versions is the coil L1.

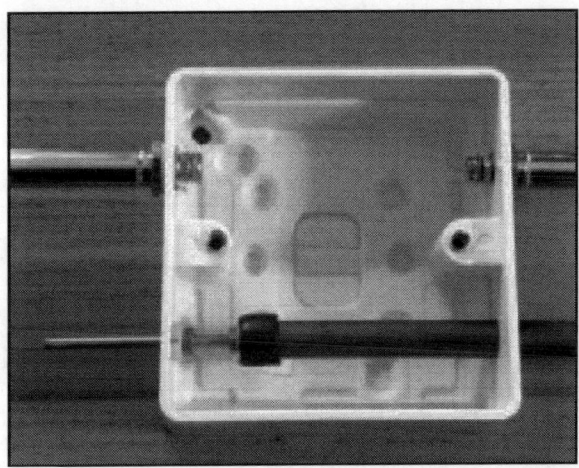

Photo 2: Mounting arrangements for the ferrite rod and dipole elements.

The coil for higher frequencies has 14 turns tapped 4 from the end, wound on a 2.5cm plastic tube. The lower frequency coil had 30 turns, tapped 10 from the end. The series capacitor Ct can be a pre-set trimmer or a small variable (eg Maplin FT78K) mounted inside the box. In the lower frequency version it may be beneficial to add a 100p capacitor in parallel with the trimmer for the lower frequency bands (see **Table 1**).

In a later development of the high frequency version I used thicker wire, oval in cross section and equivalent to about 16 SWG. Twenty turns will resonate at 14, 18, 21 and 24MHz. If the coupling winding is placed a few mm to the right of the loading coil (ie the end first entered by the ferrite rod on its insertion) it is possible to achieve a 1:1 match with no series capacitor. A further refinement, leading to only a small increase in size, was to build the unit into a 'Double Mounting Box' with the coil and tuning mechanism in the lower compartment. This had the advantage that there was negligible 'hand capacitance' effect when making adjustments to the tuning.

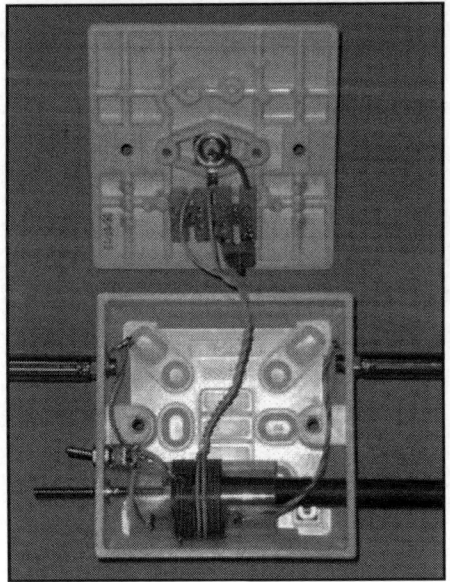

Photo 2: Mounting arrangements for the ferrite rod and dipole elements.

ASSEMBLY

In place of the original pointers I used the Maplin 10 section telescopic (1.3m) antennas (code LB10), which have a hole in the base tapped to take an M4 bolt. Most of the hardware came from B&Q and Homebase. A major 'find' was the 'Single Mounting Box' made from a synthetic plastic material for domestic electrical use. It comes 25 or 30mm deep and has sheer rectangular sides. With suitable precautions it can be hand drilled, and might have been expressly designed for the VISTA antenna.

In the past, the action of a lipstick cylinder has been used as a way of moving inductor cores, but it brings its own problems. My wife, who has no particular fondness for either lipstick or amateur radio, suggested the mechanism in **Photo 2**. This turned out to be ideal. A 40mm long, 3.5mm diameter machine screw is threaded through a cup formed from a piece of graduated rubberised cable protector. The screw thread then passes through a nut secured with cyanoacrylate adhesive (or, better still, a stripped-down and slightly crimped stereo jack socket) in the case wall, so that the ferrite rod is held firmly. Its other end protrudes through a hole drilled in the opposite side of the box as shown in the figure. The rod can be moved in and out by rotating the screw or the rod itself, allowing precise and stable tuning.

OPERATION

The VISTA antenna will tune to give a 1:1 50Ω match on any of the HF bands in its range. I use 4m of RG174 to connect it to the power meter and transceiver. In use, the antenna is set up on a suitable support and peaked for maximum receiver noise level. It is then adjusted for zero reverse power on transmit. The series capacitor can be pre-set to about 80-100pF for 14MHz, a bit more for 7MHz and much less for the higher bands. It need only be adjusted once for each band. Antenna resonance is affected by proximity to the operator, so it is necessary to tune in increments and check several times. If peaked for the middle of the 20m CW section it is possible to operate from 14.000 to 14.060MHz without retuning. Alterations to the length of the coax feed made no difference to antenna resonance. A 4m length was enough to site the antenna clear of obstructions but within easy reach to make adjustments.

Remember, this is a QRP-only aerial. Do not attempt to drive it with more than 5W.

TABLE 1: Component values for lower- and higher-frequency versions.

	LF version	HF version
Ct (nominal)	80pF	60pF
Additional capacitor	100pF none	
L1	10 turns	14 turns
Tap L1 at	10 turns	4 turns
L2	2 turns	2 turns

RESULTS

With the VISTA in a downstairs room I have had six brief CW contest exchanges with east coast USA and Canadian stations. One of these was with 2W and the others between 3.5 and 5W. Two more US contacts were made with the antenna sited in the roof space and 4.5W. Curiously, I didn't get the impression that the increased height and 10-12m of extra feed line made any difference to my footprint in the USA. Aside from contests, there have been several contacts round Europe lasting for 5 minutes or more and limited as much by my indifferent Morse as by the signal strengths.

CONCLUSION

This is another small antenna that works. Limitations? It is not an efficient radiator, though this would be improved if its electrical length were increased, perhaps with capacity hats at the ends. It does need to be retuned if its height above the ground is altered, but you do need to check the reflected power during operation. I should also mention that my home station is on a sandstone ridge between the Mersey and Dee estuaries so it has some natural advantages.

What is new? Well, it is quick and unobtrusive to set up and it fits not just in a briefcase, but in a sponge bag if you dismantle it. As it works indoors it might do even better on top of a small mountain.

REFERENCES

[1] No Cost ATU, Tony Haas, G4LDY, Sprat 28 Autumn 1987

[2] Bi-directional Wattmeter, David Stockton, GM4ZNQ, Sprat 61

The Slim Jim 2m Antenna

The Slim Jim antenna was conceived by F C Judd, G2BCX many years ago. He described details of a 2 metre version, which was reproduced in the sixth edition of the RSGB Communications Handbook. He mentioned that scaled-up ones for the HF bands could be used, but gave no details, so some practical information is given here for two other bands.

The Slim Jim antenna was conceived by F C Judd, G2BCX many years ago. He described details of a 2 metre version, which was reproduced in the sixth edition of the *RSGB Communications Handbook*. He mentioned that scaled-up ones for the HF bands could be used, but gave no details, so some practical information is given here for two other bands.

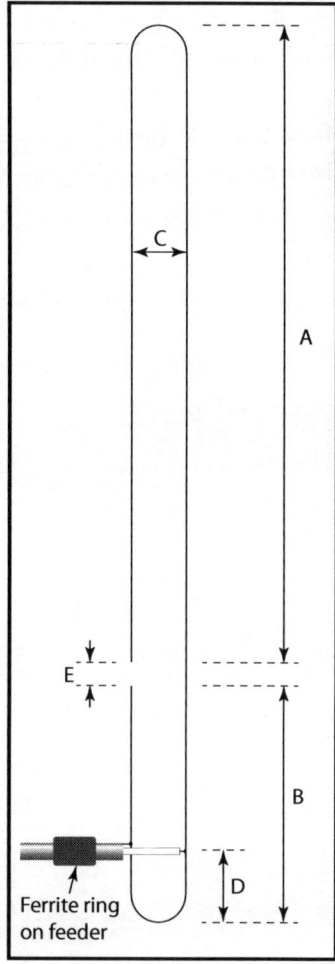

The antenna can be described as an end-fed folded dipole with a quarter-wave stub for loading. The folded dipole currents are in phase and radiating, whilst stub currents are in anti-phase and therefore not radiating. The Slim Jim has a good low-angle radiation pattern, is broad banded, has low wind resistance and is light – and inexpensive. The coax connections are low down so there is no interaction between the feeder and antenna. Adjustment for a perfect SWR is simple and very effective. Don't settle for anything worse than 1.1:1! It is a highly recommended antenna. The name? 'Jim' from the 'J' matching stub and 'Slim' because it is. **Figure 1** shows the general arrangement of a Slim Jim and **Table 1** shows dimensions for 2m, 10m and 15m. Other bands can be calculated *pro-rata*.

The antenna is relatively broad band, meaning there is no need for an ATU. Although the antenna seems to function quite happily without a balun, good practice suggests one should be used. For VHF, a ferrite ring on the coax will suffice. A ferrite might be adequate at lower frequencies but a coil of coax is probably a better solution. Omitting a balun will lead to currents flowing on the outer of the coax, which at the very least will upset the radiation pattern.

Fig 1: General arrangement of the Slim Jim antenna. For dimensions, see Table 1.

A Slim Jim for 2m

The author made the simple 2 metre Slim Jim antenna shown in **Figure 2** in a couple of hours. Using the dimensions in **Table 1**, a length of 8mm copper pipe was bent. An overlap was left between the ends, which were then trimmed to provide the correct lengths and gap. This is much easier than trying to get all the dimensions perfect at the same time before bending. There is no need to worry too much about distorting the tube at the top and bottom, except possibly for looks; but no one will notice when the antenna is high up.

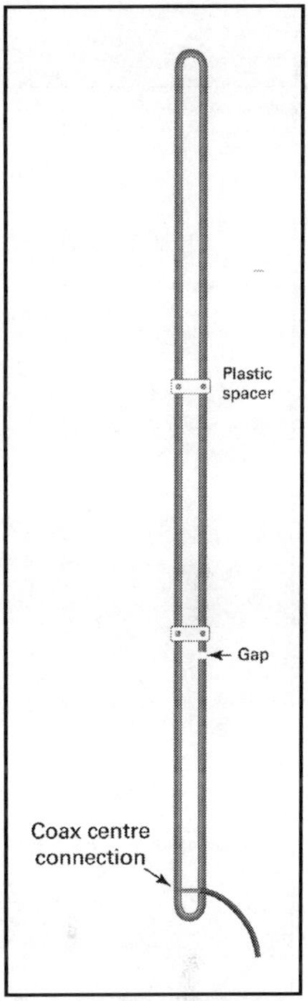

The gap at B can be filled and supported by cadging a 5mm diameter plastic knitting needle and cutting about 10cm off it, then passing it up into the two ends of the pipe, gluing it in position, with half above and half below the gap. (Mark the centre of the piece of needle before inserting it into the tube, as an aid to getting it centralised.)

The insulating spacers can be made of any suitable plastic you have available. The pieces in this one were fixed by drilling holes through the plastic and self-tapping screws used to fix it to the copper, drilling only into the first layer of the metal, and using screws that did not penetrate right across the tubing, to avoid reducing its strength.

Using the dimensions shown, the VSWR probably won't be too bad but it is best to optimise it. I only had to adjust mine by about 6mm (1/4") from the dimension D in the table to obtain a SWR of 1.05:1. The setting depends to a small extent on where the antenna is mounted and the presence of any metallic objects. It may be hard to keep the aerial upright and far enough off the ground and yet be able to make the necessary adjustments to the feed point. One trick is to suspend the antenna horizontally, about 2m above ground, and make the adjustments there. It won't be too far off when you later put the aerial at the top of your pole.

Before installing the antenna, it is important to waterproof the coax at the feed point. Traditionally, self-amalgamating tape is used, but it is also possible to use a small waterproof box provided it is properly sealed.

Mounting the antenna is not difficult and again it is up to individual preferences. The main things to consider are the local geography and available materials. It could be mounted inside a piece of plastic tubing, but that would increase windage.

FIG 2: General view of the 2m Slim Jim.

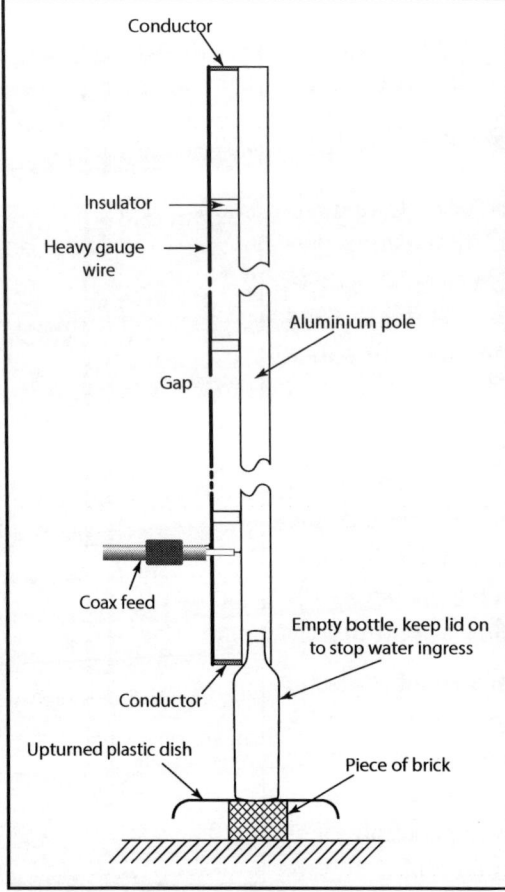

Fig 3: 21 and 28MHz Slim Jim antennas. The same basic arrangement could be used for other frequencies.

Another trick is to mount the bottom third in a length of plastic tube fixed to the top of a pole (thus keeping the bottom of the aerial clear of the pole). Experience has shown that provided the bottom third is supported, only very severe gales cause any slight bending of the copper. If you have a suitable suspension point available, eg in a loft, it could also be suspended from a line; it weighs only about 400g (under a pound) and the feeder would be about the same weight.

A 28MHz version

Also the author built a Slim Jim for 28MHz in the 1960s and used it with great success. It was mounted on a thick bamboo pole obtained from a carpet shop, which had been in the centre of a wide roll of carpet. My home was in a very rural spot and we had a hedge fifteen feet (5m) high, substantial enough to support a prop for the bamboo pole. The antenna was made of heavy gauge plastic-covered wire, but otherwise not much care was taken regarding insulation. That particular antenna had a surprisingly low feed point, but the reason was never really investigated as the SWR was very close to 1:1, so there was little incentive to probe further. The log book of the time shows many DX contacts with that inexpensive setup, using only 40W CW.

A 21MHz Slim Jim

Increasing dimensions at lower frequencies necessitates alternative construction techniques. **Figure 3** illustrates how I made one from an aluminium pole 10.2 metres (33' 2") long with a heavy-gauge copper wire running down an inch (25mm) away from the pole, mounted on insulators about every foot (30cm). The pole forms the 'spine' of the Slim Jim and provides support; wire is used for the rest of the aerial (lengths A and B in **Figure 1**). The bottom of the pole can have a high voltage on it, so it was insulated by mounting it on a bottle. There can be quite a lot of mechanical stress on the bottle so I recommend you use one made from fairly thick glass. An old plastic dish turned upside down protected the space underneath, where support was provided by a brick. use 'Chemical Metal' from Plastic Padding to glue these parts together. It provides sealing and mechanical strength.

Make sure that connections to the aluminium pole are sound because oxide forms very quickly, producing a very thin insulating layer. Scrubbing with emery cloth covered with a little grease can be useful. The emery breaks down the existing oxide film and the grease stops oxygen getting to the aluminium, preventing the oxide reforming. Smear a little grease on the cleaned surfaces before clamping up.

Further uses for the 21MHz Slim Jim pole

As the top of the antenna is about 35 feet (11 metres) up, it can be a useful 'skyhook'. A line passed up and through a ring at the top and taken to a high point such as a house or tree, and through another ring, allows for the line to be dropped down and raised up again at both ends. Pulleys could be used, but experience has shown that they are likely to cause jamming. Rings are simple – and cheap. Vertical antennas can be suspended at intervals along the line, using strimmer plastic for the first foot or two.

For a simple mast of this height, the supports do not have to be very substantial. Three guys of 4mm nylon line at roughly 120° spacing taken out as far as possible from the base of the mast are fine. If the pole is well supported at least ten feet (3m) above the ground and a sturdy insulated bracket holds the base just above the bottle it would not need any guys, much as a flagpole does not require one. If a line is used as suggested above for other wire antennas, one guy in the opposite direction would counteract the pull of the line.

Conclusions

Although a low radiation angle is not always very useful for local contacts, it *is* good for DX working and the Slim Jim performs very well, considering its simplicity and ease of construction. Experience shows that it is as good as a full-blown ground plane antenna, takes up little space, is robust, and, very much in its favour, inexpensive.

TABLE 1: Slim Jim dimensions for different bands (see Figure 1). Lengths are given in mm followed by (feet and inches).

Dimension	145MHz	28.8MHz	21.1MHz
A	990 (3' 3")	5040 (16' 6")	6800 (22' 4")
B	495 (1' 7.5")	2520 (8'2")	3400 (11' 2")
C	45 (1.75")	45 (1.75")	45 (1.75")
D*	102 (4")	530 (1' 9")	700 (2' 4")
E	2504 (1")	25.4 (1")	25.4 (1")

* Approximate – move the feeder connections up and down for a perfect SWR.

The basic design of the antenna is nothing special – just a long wire – but the clever bit is how to keep it up in the air, tree's are ideal for supporting one end of an aerial wire. But how to get the wire up there without the assistance of a trained squirrel? Fortunately,

The basic principle is to use a fishing rod and reel to loft a weighted, non conductive leader line to the top of a tree. The weight should cause the line to hook over a branch and head for ground level. The other end of the line is permanently attached to a 132 foot (40.2m) length of wire that acts as the aerial element. A stand for the fishing rod plus a couple of guys and some wiring completes the ensemble, as shown in **Figure 1**.

The parts are basic and quite easy to obtain. When the antenna is dismantled it will fit in a car for easy portability, ready for use at any time. The only really long piece is the aluminium support tube, which could be cut in half and then sleeve jointed. I find it takes less than ten minutes to erect once a suitable tree has been selected.

PARTS AND CONSTRUCTION

None of the parts are terribly critical, so just use the following as a guide. You will need a beach caster type fishing rod. I used a telescopic one, but any kind will do as long as it is about eight or ten feet long. Avoid the

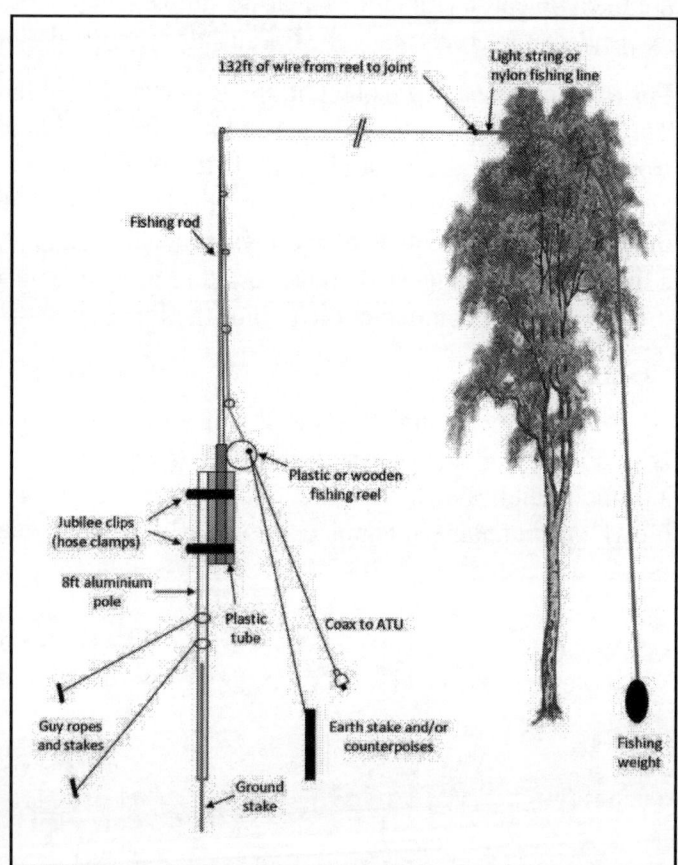

Fig 1: General arrangement of the fishing rod aerial system.

conductive carbon fibre types, though. You will also need a centre pin type plastic or wooden fishing reel. I made my own from a spool that once contained a lot of enamelled copper wire, but a commercial fly reel will be fine.

Make a connection point in the side of the reel – I just put a bolt through. You may find it convenient to remove the reel winder knob and put a bolt through in its place. Use a solder tag to connect one end of a 132ft piece of thin wire. I used 22 SWG stainless steel stranded insulated wire, although this is not ideal because of its high resistance. You can use any wire you like as long as it is strong enough and thin enough to fit comfortably on the spool with space to spare. Next, attach a long 'leader' of string or nylon fishing wire to the other end of the wire. About 60 feet (20m) is enough. A solder tag at the end of the wire provides a handy attachment point. Wind the string onto the reel on top of the wire. **Photo 2** shows the general construction.

The rod support is constructed from an 8 foot or so (~2.5m) length of aluminium tube, of roughly 1.5" (37mm) diameter. Dimensions are not at all critical. Attached to the support is a piece of PVC pipe of a suitable diameter to take the bottom of the fishing rod. The pipe is about 16 inches (40cm) long and doesn't have to be a tight fit to the rod. I used two jubilee clips (hose clamps) to attach the plastic pipe to the support rod. **Photo 3** shows how it all goes together.

Photo 2: How I terminated the wire on the bolt and (inset, right) the bracket I made to fix my homebrew reel to the rod.

TABLE 1: Suggested counterpoise lengths	
Band	Length
160m	39.5m / 129'6"
80m 2	0.5m / 67'5"
40m	10.5m / 34'8"
30m	7.4m / 24'4"
20m	5.3m / 17'4"
17m	4.1m / 13'7"
15m	3.5m / 11'7"
12m	3m / 9/10"
10m	2.7m / 8'9"

Photo 1: General view up the fishing pole showing the support pole, sleeve, reel and rod.

Photo 3: General construction of the rod support. Note how the tube protrudes an inch or so above the support pole.

Photo 4: Encouraging the ground stake to stay vertical.

I used a piece of steel reinforcing bar about four or five feet long as my ground stake. A slight point on the end will help it go into the ground, as will encouragement with a suitable implement (**Photo 4**).

The final parts to make are the feed and counterpoise. My prototype used a 10' (3m) length of RG58 coax with a PL259 plug on one end to suit my ATU. The other end has an alligator clip on the centre conductor to connect to the end of the aerial wire (**Photo 5**). The counterpoise length is calculated as 75/frequency (MHz), which allows a bit of extra length for trimming. **Table 1** gives suggested values for the mid-point of the HF bands, though you may well find that trimming these by 5% or so will be better. I only use a single counterpoise wire per band, although I recognise that more might be better.

DEPLOYMENT

This description assumes that you know how to beach cast a fishing rod. If you don't have the knack then please find someone to teach you otherwise you could injure yourself or others. Select a suitable tree and make sure that there are no people or animals nearby that could be hurt when you cast the leader. Trees beside footpaths are particularly prone to people walking near them, and folks tend to get upset if you hit them with flying lead. Respect the wildlife that may be in the tree -after all it's their home!

Thread the leader through the rod loops (just like a fishing line) and attach the weight to the end of the leader. I let a goodly bit of slack off the reel, ensuring it doesn't tangle. I think it's called 'flaking out' the line. Don't try to cast straight off the reel or a 'bird's nest' (tangle) will result. Beach cast toward the top of the tree. With luck the weight will carry the leader over a high branch and fall to the ground. Pull the leader over the branch so that the

Photo 5: Attaching the feed (the antenna was not deployed at this point, which is why the antenna wire is not visible).

end of the aerial wire is several feet from the leaf canopy. Tie off the leader at the base of the tree.

Go back to the rod and pay out the aerial wire as you walk away from the tree. When the wire is fully extended, set up the ground stake, slip the rod support over it and then put the rod in the top of the support. If the aerial wire is a bit saggy then you can go back to the tree and tighten it by pulling on the leader.

Depending on the stoutness of your ground stake and the weight of your aerial wire, you may find it necessary to use some guys to keep the rod support upright.

Finally, connect the feed to the bolt on the reel and arrange your counterpoises.

IN USE

The radiation pattern will mostly be to the sides as a figure of 8. However, if you select a very tall tree you'll find that it acts as a reflector and the radiation pattern approaches that which you'd get from a sloper attached to a tower.

There are other methods of feeding the aerial. One that may be attractive is to put an automatic ATU on the ground at the base of the aerial, with a single wire connecting to the driven element. The ATU earth can then be connected to counterpoise(s) or even just to an earth stake. I use a small LDG auto tuner and it has always served me very well.

I have used this aerial for years with a Yaesu FT-840, an SWR meter and my trusty LDG auto ATU. If it's windy then the SWR will alarm you as the tree sways about but I've not had any real problems in practice. Happy DXing – or should I say "tight lines"?

Believe it or not, a wide range aerial made from a child's toy, this version comprises 32 turns of an original Slinky, 2042 mm of 1/2" plastic water pipe, a plywood base, strong string and, most importantly, one 500pF variable capacitor.

INTRODUCTION

If I told you I made an antenna 650mm in diameter that can transmit 80 to 17m with 100 watts and a SWR of less than 1.2:1, you would tell me that I am smoking my socks. Well, let me tell you the story.

We, that is my XYL and I, retired in South Africa where antennas are generally big, then relocated to my son's place in Lindfield, West Sussex. We have now moved in to a retirement flat in Haywards Heath, with no ground and no antennas allowed – and a noise level of S9+20. I was ready to take up tiddlywinks.

I asked around at the club and was given a loop design that jogged my memory. In the November 2003 *Practical Wireless* John Heys, G3BDQ, wrote about a Slinky Hula antenna he made with a Slinky, a hula hoop and a 120pF capacitor. So, using his design and a bit of my mechanical skills I have made my own version. It comprises 32 turns of an original Slinky, 2042 mm of 1/2" plastic water pipe, a plywood base, strong string and, most importantly, one 500pF variable capacitor

Now, I just happened to have a Jennings vacuum capacitor, obtained from a boot sale in South Africa. This is the Rolls of capacitors, not for the faint hearted when buying a new one. I am also told that even if it has lost its vacuum you can still transmit up to 100W. If you are going to experiment with standard variable caps I would suggest a slow-motion

Photo 1: The completed Slinky loop antenna, mounted on a tripod.

drive. I found a 120pF did not tune down as low or as high as the 500pF cap and was very sensitive to the touch.

CONSTRUCTION

I started off with the pipe, marked the centre, 29mm either side of the centre and then at 58mm intervals 15 times either side of the centre. I then drilled 5mm holes right through the pipe (in a straight line).

Photo 2: Detail of how the Slinky is mounted on the water pipe. The string is tied to the Slinky on the hidden side of the pipe.

The Slinky is available from several sources. Mine came from Maplin in Brighton. Count 32 rings of the Slinky and cut it with a strong pair of wire snips. Screw one side of the pipe to a suitable or plastic base, making sure it is in the same plane as the rest of the holes. Thread the Slinky over the pipe, bend the pipe round and then screw the other side of the pipe to the base.

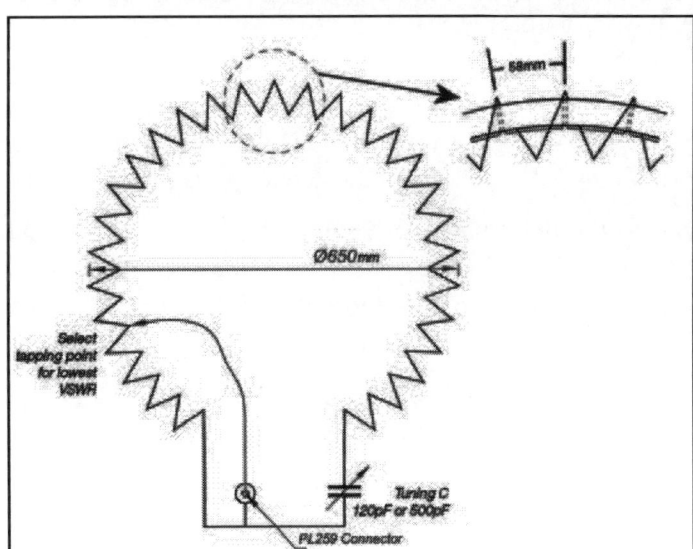

Fig 1: Circuit of the Slinky loop antenna. See text for details.

Start with tying one full ring of the Slinky to the pipe and then one on each hole, ending with a full ring again. **Figure 1** and **Photo 2** show the basic arrangement.

I made up a bracket to hold the input socket and vacuum capacitor (see **Photo 3**). It doesn't need to be anything special, but there must be a good connection between the body of the input socket and the earthy side of the capacitor.

SETTING UP

It's important to find the right tapping point for best VSWR. To start with, connect the centre pin of the input socket about a quarter of the way round the loop. Tune the loop to resonance (listen on your receiver and adjust for maximum noise). Then transmit at low power, measuring the SWR. You will probably have to move the tapping point several times and maybe re-tune the capacitor because the settings can

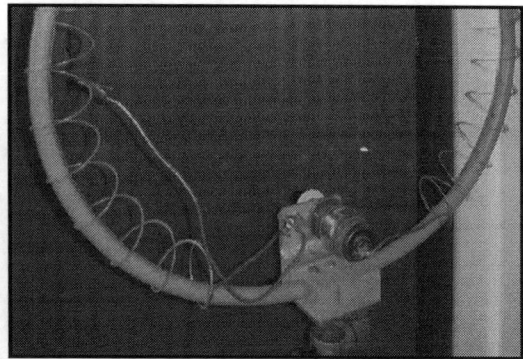

Photo 3: *General view of the business end of the loop, showing the mounting block, capacitor/input socket bracket, wiring and my tapping point.*

interact. However, once you've got a good match, you shouldn't have to change the tap for different bands.

I used my MFJ analyser and found I could get a 1.2:1 SWR anywhere between 80m and 17m.

IN USE

Initial tests on 5MHz by Ken, G3WYN got very good results. The loop was only a couple of S-points down on his full sized dipole, which surprised us.

As we have just moved into the flat I have not been able to experiment much further. But I am going to continue playing because it would be nice to be able to tune it remotely. I would like some one to design (cheaply!) a stop for each end of the vacuum capacitor drive if I use a stepper motor.

Have a go at building this antenna. I would like to know how you guys get on, maybe we all can benefit from your experiments too.

Tom Haylock, M0ZSA

The Selfie-Stick Yagi

Many people have one of those selfie-stick things hanging around. Well, I have found a much cooler use than taking blurry I'm-the-centre-of-the-universe photos: turn it into an amateur radio aerial.

HISTORY

I have been experimenting with small (ish) direction-al antennas for a while for my summit and portable operating. Two years ago, I tried a roll-up Yagi. This was achieved by mounting the elements on a length of webbing tape. A gamma match was employed for tuning. I also found that it was possible to mount a 144MHz and 432MHz

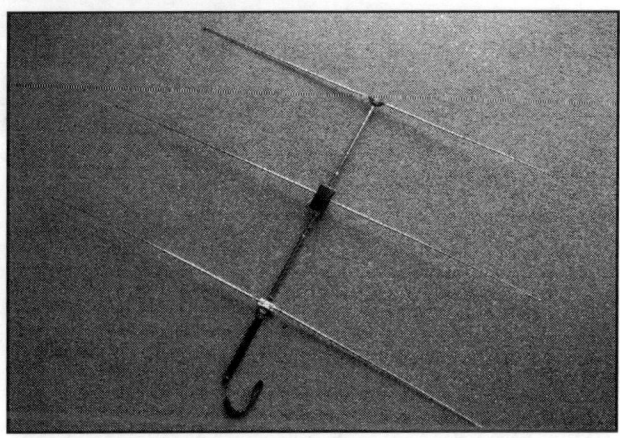

Photo 1: The prototype selfie-stick antenna.

beam on the same piece of tape and employ one feed to a pair of gamma matches, thereby making a rather useful 4-element VHF and 8-ele UHF antenna. To deploy it, the elements were attached to a walking pole with releasable zip ties. A small bracket bolted to the walking pole allowed it to be mounted horizontally or verti-

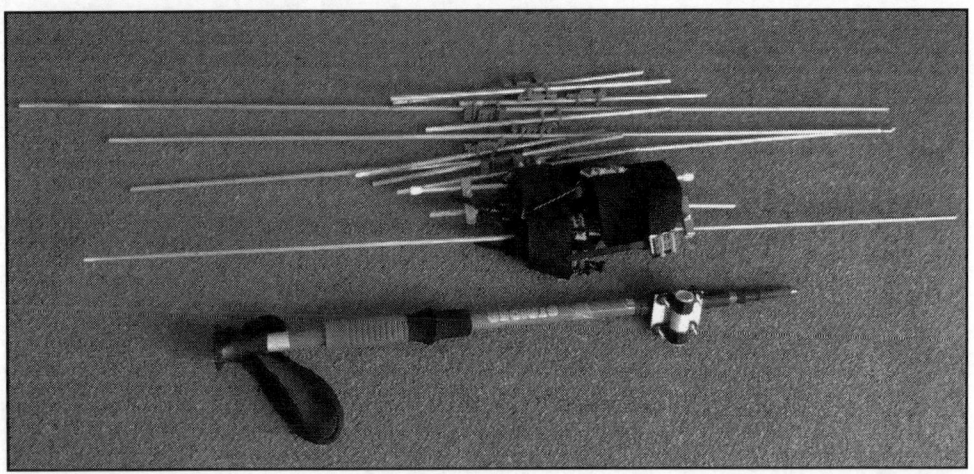

Photo 2: Earlier roll-up dual band Yagi, ready for transport. The walking pole in the foreground shows the mounting bracket.

cally on a Photo 2 and Photo 3 show how this worked.

Some very good contacts were made with this antenna. The elements were made from old archery arrows, which made for a robust design when deployed, but it was a little time consuming to set up.

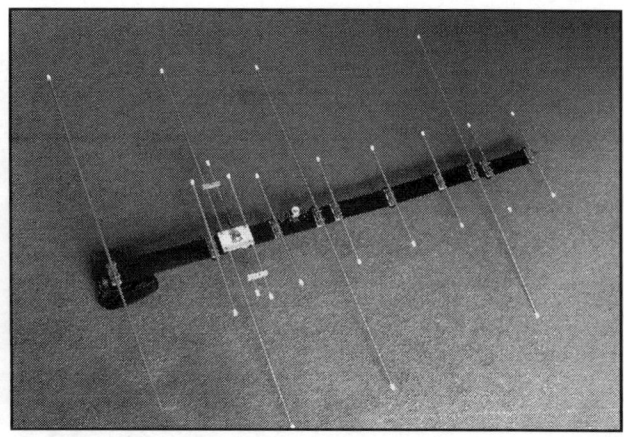

Photo 3: Roll-up Yagi attached to a walking pole and ready for use.

My next 2m antenna was a simple Moxon. It used lightweight plastic pipes for the spreaders. This collapses down nicely for transport and is very quick to deploy. It became our antenna of choice on lightweight activations, such as that shown in **Photo 4**.

But, what I really wanted was something for the occasional activation and to pick up signals from the ISS. Ideally this would be a beam that fitted in a pocket of my rucksack.

SELFIE-STICKS

In case you haven't come across selfie-sticks, they have much in common with telescopic antenna rods found, for example, on

Photo 4: Lightweight Moxon mounted on our guyed fibreglass fishing rod atop Red Screes.

broadcast FM portable radios (remember them?). My daughter, Lauren, M6HLR had a couple of selfie-sticks given to her at Christmas and she gave one to me and I wondered if I could do anything constructive with it. The length looked useful so I started to see if it was possible to make a beam antenna with one. **Photo 1** shows the result and I'll describe the various steps I took to create it. Note, however, that the dimensions I've listed here are all specific to the bits and pieces I used. The materials you find to build yours will be different and so will be the exact dimensions. It's easy enough to find Yagi dimension calculators online and plug in the sizes of whatever you've got to hand. Go with what the calculator says: only worry

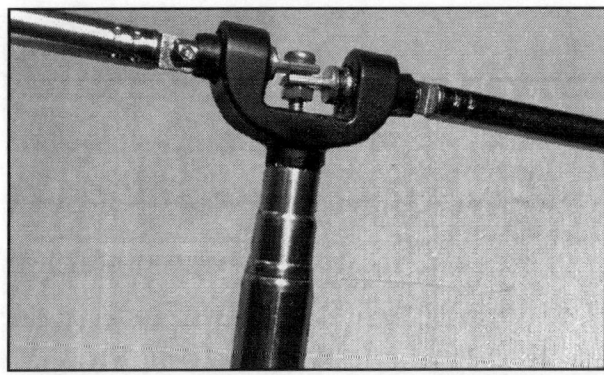

Photo 5: C-shaped bracket adapted to mount the reflector.

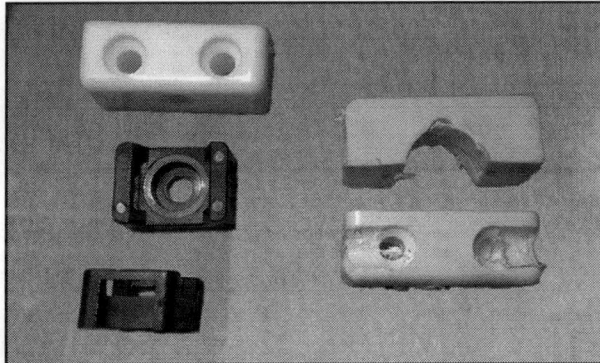

Photo 6: Modesty blocks after drilling.

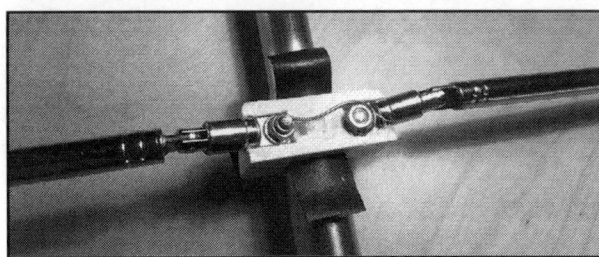

Photo 7: Finished director centre.

if it gives you lengths markedly different from mine (eg a reflector of about 1500mm instead of about 1000mm).

Figure 1 shows the general layout and dimensions of my prototype.

I purchased 6 x 50cm telescopic whips for the elements (making 7 in total), intending to make a 3-element Yagi. The rest of the components came from the shack and DIY stuff in the garage. Nothing is particularly special and you can merrily substitute items – just remember to put the dimensions into the online calculator.

The selfie-sticks each came with a bracket to hold the phone. When I removed that, it left a C-shaped brace, ideal for mounting the director (frontmost) element. I carefully drilled out the bolt so that two whip elements could be slotted through and bolted together with an M3 nut and bolt. I found that drilling a small hole directly down the middle of the bracket allowed the M3 bolt to slot in, thereby stopping the element from twisting. **Photo 5** shows the result.

My attention then went to the director (the rearmost element). To make the mount, I used furniture (or 'modesty') blocks – those little plastic widgets that let you attach sheets of timber at right angles to each other.

The first step was to cut two grooves for the whips. This was done by clamping two blocks back to back and simply drilling down to the fixing hole. The same concept was employed to make the larger curved groove that attaches the bracket to the selfie-stick. To make the retaining clamp, I used a plastic strip from a zip tie and drilled it with two 3mm holes. Photo 6 shows the blocks just after drilling and Photo 7 the finished director centre. You will notice that the two whips are not joined. I used a bridging wire to connect the elements together, thereby increasing the overall element length by 20mm.

Photo 8: *Dipole centre.*

Now on to the driven element. I used a small project box, a bulkhead BNC socket and another modesty block to form the centre. I drilled into the narrow ends of the modesty block and then cut in half lengthways. This left the slots for the elements. Then it was just a matter of drilling a couple of holes in the plastic box for the elements and a bigger hole for the BNC socket.

Connections between the BNC socket and elements were made using a couple of M3 solder tags attached to small bit of coax. The half section of modesty block was then secured to the plastic box with the same M3 screws that hold the driven element in place. **Photo 8** shows the finished thing.

I used a couple of zip tie wall mounts bolted to the underside of the box, which made it easy to zip tie the driven element assembly to the selfie-stick.

TESTING

Using dimensions from whichever Yagi calculator you used, extend the selfie-stick and secure the driven element the correct distance from the reflector. You may find a strip of rubber tape stops the element from sliding on the selfie-stick boom.

I found that when fully extended, the director and driven element were more or

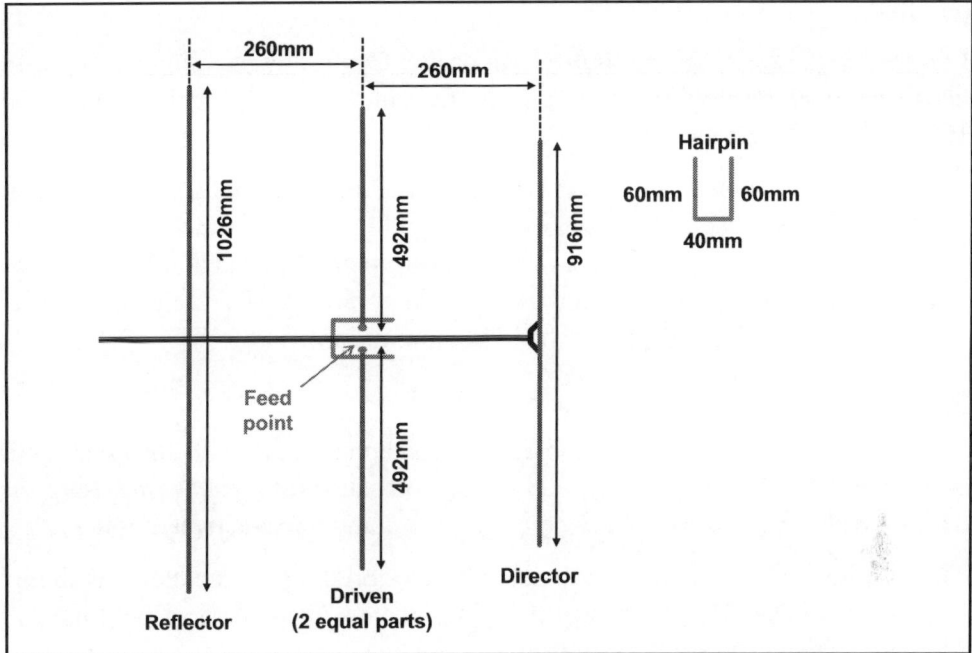

Fig 1: *General assembly of the selfie-stick Yagi. Note that these dimensions are for guidance only – see text.*

less the correct length. The director was a bit long, so I retracted the the thinnest section.

To tune the aerial, I used a piece of stainless steel wire left over from another project to make a hairpin match (but any other similar piece of solid wire will do). This was bent into a C shape and slotted into the gaps of the hinge section on the two driven element whips. Experimenting with the length of the hairpin, we were able to tune the antenna. **Figure 1** and **Photo 8** will probably help you see what I mean.

Photo 9: *Selfie-stick Yagi in use, wielded by Lauren, M6HLR.*

IN USE

I first tested the aerial during a QRP Backpacker Contest by mouning it on our 4m telescopic pole. Perhaps it isn't the most efficient contesting antenna but several 100+ mile contacts were made.

The biggest issue when using the antenna like this is the wind. I constantly had to straighten the elements, as the hinges would fold in the wind. There are probably several solutions for this. I used a couple of biro pen tubes and a piece of insulation tape across the element hinges, which temporarily stiffened them, however this isn't really an issue if you are hand holding the antenna.

FINALLY

The end result may not be the most robust antenna around but, for quick activations and portability, you will find it hard to beat. **Photo 9** show me using the aerial – with M6HLR acting as a combined mast and voice-activated rotator.

This is a fun and simple project to make. With a bit of experimentation it should be possible to use the same basic method to create a 70cm version, but I haven't tried it yet. Naturally, all the dimensions will be about one-third those for the 2m version, so the selfie-stick telescopic rods may not be the best choice. However, let the online Yagi calculators be your guide.

Kevin Richardson, G0PEK

avckev@btopenworld.com

A Frame Aerial For HF

A receive-only aerial does not have to be efficient to be effective. The most important attribute for receiving is the signal-to-noise ratio. Because the noise source is usually localised, a directional aerial can often be used to remove noise, leaving the required signal in the clear.

THE PROBLEM

Tuned ferrite rod aerials can be very good at achieving this, but I have always found that on HF they are not as effective as they are on the LF bands. The frame aerial is far superior and you do not have to find suitable ferrite!

THE SOLUTION

To achieve a wide tuning range, the usual solution is to have a large coil of wire and select sections of it using a switch. A variable capacitor is then used to resonate the coil on the band in use.

My solution was to make the loop resonate at several frequencies at the same time. The circuit diagram is shown in **Fig 1**. C1, C2 and C3 are the three gangs of a 500pF tuning capacitor, as found in older broadcast receivers and at rallies. L1, L2 and L3 are sections of a single-frame coil as shown in **Fig 2** and tapped at two points. The circuit's behaviour is very complex, with at least five resonant frequencies for any given setting of the capacitor, but the three main ones cover the amateur bands.

The tuned circuit formed by L3, C2 and C3 is designed to cover the range 11 to 30MHz for a full sweep of the tuning capacitor. L2 with C1 and C2 resonate at about 7 to 15MHz, and L1 with C1, C2 and C3 cover 1.5 to 4MHz. It must be expected that even a slight variation in design will vary these bands, but the differ-

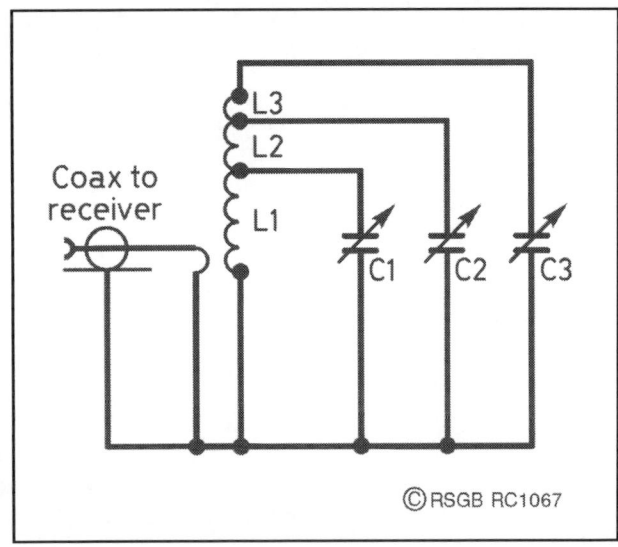

Fig 1. *HF frame aerial circuit diagram.*

ences between my Mk1 and Mk2 efforts were slight, the main difference being that Mk1 worked up to 35MHz, whereas the final one just reaches 30MHz.

Coupling the aerial to the receiver is accomplished by a single-turn loop round the outside of the frame aerial.

CONSTRUCTION

This is up to the constructor to a large degree, but I used 15mm square timber, half-slotted to make the cross. I then used six 10mm pins hammered in at the

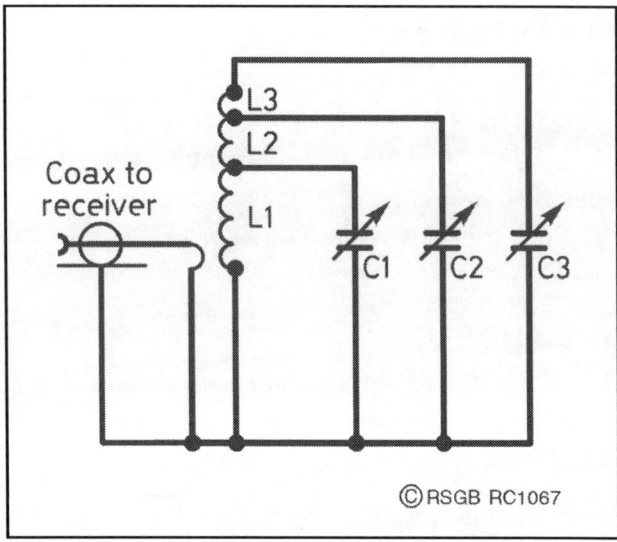

Fig 2. Construction of the frame aerial.

ends of three of the arms, leaving 3mm exposed. The fourth and lowest arm needs seven pins. These pins were used to support the tinned copper wire, which was then wound round the pins and soldered to hold it in place, as **Fig 2** illustrates. The three-gang capacitor was fixed to the base plate and short lengths of wire used to connect to the frame aerial.

Portable Magnetic Loop

For many years, a portable magnetic loop has been in use at G3HBN for holidays and special events. The design was a simple loop of RG-213 braiding slid over a piece of half-inch water hosepipe and supported by pieces of bamboo which formed a pear-shape. The whole was manually tuned and supported on a photographic tripod. It was time to upgrade this loop. Some articles in RadCom inspired a new design.

The requirement was to improve the performance and increase the operating bandwidth, if possible. Looking at some aerial history, the old Cage Dipole came to mind. The Cage Dipole was designed to increase the bandwidth and help with the matching of it for commercial broadcasting purposes. In those stations, very long open-wire feeders from the transmitter to the aerial were customary. Aerial tuning units were not used and the feeder was coupled directly into the transmitter with either a link or a π-coupling circuit. Such aerials would have a bandwidth of say 2.5 to 5.0 or 5.0 to 10.0MHz.

THE LOOP

The element of this loop is constructed along the lines of the cage dipole outlined above, but only a single cage element is used. This element consists of 12 cables of stranded plastic-covered hook-up wire connected in parallel. The inner core of wire is about 1mm diameter. The overall diameter of the cage is 50mm. The element, when constructed, is placed on an hexagonal wooden frame. A ganged tuning capacitor of

Photo 1: The loop in use at G3HBN.

525 + 525pF with slow-motion drive is used to bring the loop to resonance at the desired frequency. The tuning range is from 6.9 to 32.0MHz. The loop is fed with a Faraday link coupling made with RG-213 coax, the braiding of which is open for 2.5cm at the centre. The photograph shows the completed loop on a tripod mounting. One of the problems of magnetic loops is the very narrow bandwidth. **Table 1** illustrates the comparison between a 1m, 22mm copper tube element and the 1m, 50mm caged element.

Portable loop 12-wire cage	Centre Frequency	Fixed loop 22m copper tube
21	7015	11
30	10115	16
50	14050	30
75	18100	45
110	21100	60

Notes: All values in kHz.
Measurements were taken at VSWR of 1.3:1 points with a Welz SP-300 VSWR / power meter.
Both loops had a VSWR of 1:1 at the centre frequency.

Table 1. Bandwidths of the two designs.

A worthwhile bandwidth increase has been achieved. This represents an overall improvement in performance and is reflected in the results. The outer diameter of the loop at the diagonals is about 105cm. When the length of the conducting element becomes greater than $\lambda/4$, the loop ceases to operate properly and becomes difficult to couple. It is desirable, therefore, to try to keep the overall conductor length to about 0.24λ or less at the highest frequency of operation. Although the model described here will operate on 29MHz, the element is really just a little too long.

CONSTRUCTION

Most constructors seldom follow exactly what is described in an article but, listed in **Table 2**, are the items that went into making this particular model.

3	old CDs glued together to form the centre core
2	plastic 3cm plumbing nuts glued together
7	plastic till-roll spools or similar rigid plastic tube that fits the dowels
12mm	dowelling for the 6 spokes of the hexagon

5	plastic 10mm or 15mm wall-mounting water pipe clips
4	2.5cm (1in) rubber tap washers
5	3 x 40mm bolts with nuts and washers
5	water pipe saddles to strengthen centre mountings
12	55mm plastic discs.
4	packets of 10m of 6A (24/0.2mm) hook-up wire (Maplin)
	Connectors to suit termination to the capacitor
1	525pF + 525pF variable capacitor
1	suitable plastic box or container
1	slow motion drive 6 or 7:1 reduction.
1	well-insulated tuning knob, or plastic coupler and knob
1m	RG-213 Coax
5	Terry clips, 10mm
	Portable and rotary mounting methods

Table 2. Parts list.

The assembly is fairly obvious from the photographs and **Fig 1**. There are several points that are not so obvious. The overall diagonal measurements from the centre to the outer cables should not be less than about 105cm. If the hexagon is smaller than this, with the cable specified, the loop might not quite tune to 7MHz. With this measurement, the loop should tune from 6960kHz to 32MHz.

The dowelling should be cut into five lengths of 47cm and inserted into rigid plastic tubes (eg till-roll inners) at the centre hub. The sixth length is measured to fit whatever mounting box can be found. The seventh plastic support should be mounted to the tuning box and then the sixth spoke measured and cut. The five white plastic pipe clips are screwed to the ends of the five 47cm spokes.

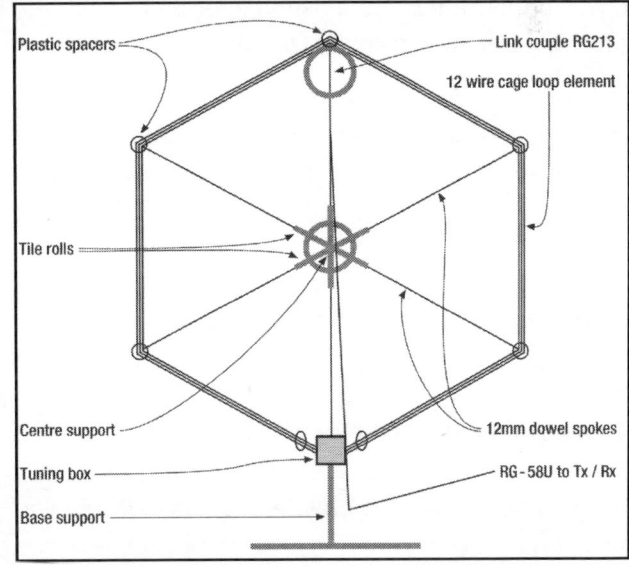

Fig 1: The magnetic loop – for dimensions, see text.

The spacers are made with the 12 plastic discs cut from about 1mm thick plastic. A 3-litre food container is cut into flat pieces which are scored with 55mm circles. A further circle of 50mm diameter is scored inside each one. Marks should be made every 30° on the 50mm circle for drilling the holes for the cables. A centre hole should be drilled for the fixing bolts and rubber washers.

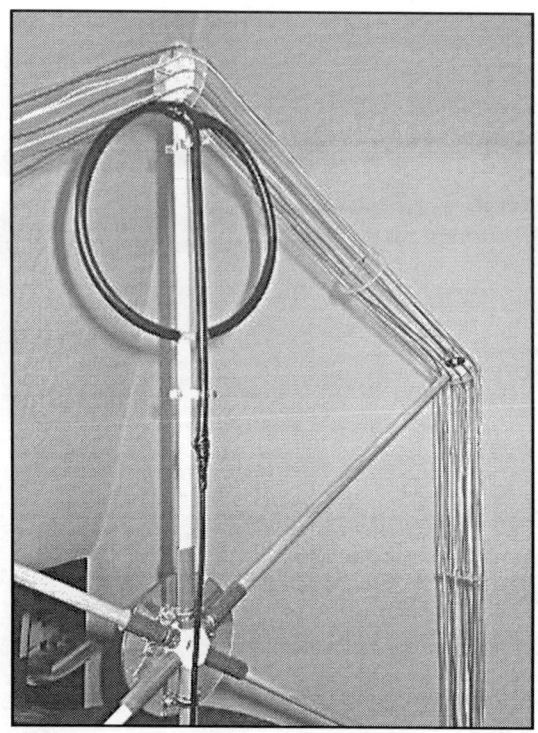

The capacitor should be mounted and some suitable terminals used. PL-259 and SO-239 plugs and sockets were used in the example, but ordinary spade terminals would be easier and just as good. The capacitor is wired to the connectors and loop using only the fixed vanes, the rotor being left unconnected or 'floating'. This halves the capacitance but doubles the working volt-

Photo 2: Detail of the loop centre and link.

age. Further, it has the advantage of there being no moving contacts involved in the RF path. A well-insulated knob, or plastic shaft coupler and knob should be used for tuning.

When the whole framework is assembled, the loop can be wired. Because the conductor is formed in a cage, the cables forming this cage will be of different lengths. Each wire must be threaded through the holes in the plastic spacers on the frame. The 12 wires can be cut roughly to length with plenty to spare for termination (coloured wires are a great help here) or from a reel of cable. But, in either case, each wire must be run separately, like stringing a musical instrument. Start with the cables nearest the centre and work outward to the top. Terminate all 12 cables at one end first and solder them to a lug or terminal or PL-259 etc. Connect this end to the tuning box. The other end is more difficult, because the wires need to be reasonably tensioned to form the cage. If they are a little slack, the rubber washers holding the spacers to the spokes provide some adjustment for this purpose. Once the loop element is finished, don't forget to mark which is the top of the cage on the plastic spacers!

Great care must be taken with all soldered connections to minimise any DC resistance.

The coupling link is made with RG-213. Many trials were conducted to optimise the coupling but, with the 'fatter' conductor for the loop element, it was found that the coupling link also needed to be 'fatter'. To form the link, strip about

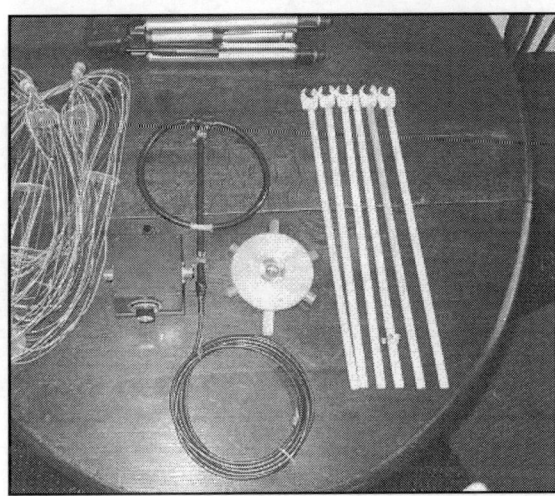

Photo 3: Ready for packing.

4cm off the cover, expose the inner conductor for 2.5cm and solder the braiding to the inner. From the tip of the join, measure 60cm of cable. Strip about 3cm of cover to expose the braiding and solder the shorted end to the exposed braiding. This will form the coupling link of about 19 to 20cm diameter. At the centre of the link, cut the braiding for about 2.5cm to expose the inner core as illustrated. Bend the remaining coax from the join to run vertically through the centre of the link and join it to the desired length of RG-58U. The mountings for the link are made with two Terry clips bolted back-to-back which clip neatly over the centre spoke and the RG-213.

OPERATION

The first step in tuning the loop is to peak it for maximum aerial noise and/or signal strength on a receiver. If an aerial analyser is available, a quick check can be made for all the bands to observe the VSWR. A 1:1 VSWR should be obtainable on all bands from 7 to 29MHz. If 1:1 cannot be achieved, the aerial should be rotated, observing the VSWR at the same time. Adjacent objects in a room can unbalance the loop and, under these circumstances, it might not be possible to obtain the necessary VSWR. A VSWR/power meter is an asset, but the loop can be tuned with a simple field strength meter, tuning for maximum signal. Remember that maximum field strength radiation will be in the plane of the loop. On receive, this will be quite marked and a null will be obtained when the aerial is broadside to the signal. However, this does not always seem to be the case. Often very strong signals are received broadside to the loop and the same signal path seems to be effective for both transmission and reception. This may be due to building reflections, another illustration of where the loop needs to be easily and quickly rotated. This is particularly useful when operating low power (QRP).

At the lower frequencies, the tuning of the loop is very sharp and it is essential to have a slow-motion drive fitted to the tuning capacitor. On the higher bands, the tuning is not quite so sharp, but it is still beneficial to tune the loop 'on the nose'. Hand-capacity has not proved to be a problem with this design. On the lower frequencies, the tuning capacitor is large and therefore any hand-capacity is insignificant. On the higher frequency bands, where hand-capacity is noticeable, it is still not too serious a problem be-

Photo 4: The tuning box

cause, at those frequencies, the usable bandwidth is much greater. There is, of course, absolutely no requirement for an ATU, if one is fitted to the equipment it should be either de-selected or tuned to a 50Ω load before using the loop.

This model is not built for high power, but it will comfortably handle 20 – 25 W. Normally, any indoor aerial used in a built-up area should not really be used for high power, since the problems of RFI (Radio Frequency Interference) can become critical. The loop should be placed as far away from the operator as is practicable.

RESULTS

Tests were carried out in CW (Morse) from the location pictured, in London, and from a seaside cottage in Folkestone, with power levels of 5W and 20W. The majority of contacts made were at the 5W level. The overall feel of the aerial was quite amazing, with signals over two S-units stronger than with the original 80cm loop, particularly on the lower-frequency bands. Comparison tests were made between the portable loop and the octagonal loop of 1m diameter on the roof. The roof aerial was, in the main, about 1 S-unit better, but the reports were that QSB (fading) was more prevalent with the indoor loop. Signal reports received varied with conditions, but reports of RST 579 and 589 were not unusual. Throughout the three-month trial period, frequent contacts were made with most European countries, Asia and North America. With 20W, the reply rate from stations called was between 70 and 80%; with 5W it was about 60 to 70% This is about the norm for QRP operation. Calling CQ was not very profitable and seldom is with QRP. Many two-way QRP contacts were also made, one notably on 30m with GM3OXX (1W RST 579), G3HBN being in Folkestone (5W RST 599), the loop being at ground level in the sitting room, about 20m above sea level. This was a very long ragchew. Conditions throughout the test period have been at an all-time low and extremely difficult for making reliable evaluations. However, several DX stations were worked with QRP and that in itself was most gratifying. No tests were made for RF feedback when using a microphone.

CONCLUSIONS

The object of improving the original loop has been achieved with greater success than expected. It seems the application of a multi-cable radiating element for the loop has brought with it more benefits than originally anticipated. The increased bandwidth is far greater than expectations and the improved overall performance in the liveliness of the aerial was a pleasant surprise. The next move is to replace the octagonal loop on the roof with a weather-proofed multi-conductor version. The results obtained here also open the door for further development in the general approach to the magnetic loop as an aerial in its own right, not necessarily to be compared with other aerial types. It is a radiator that has many characteristics that would seem not yet to have been fully exploited.

An aerial tuning unit (ATU) is very useful if you are a licensed radio amateur or a short-wave listener. The purpose of an ATU is to adjust the aerial feed impedance so that it is very close to the 50Ω impedance of the receiver or transmitter, a process known as matching. When used with a receiver, an ATU can dramatically improve the signal-to-noise ratio of the received signal. On transmit, the aerial must be matched to the transmitter so that the power amplifier operates efficiently.

DESIGN

The basic ATU design is called a T-match; you can see the basic shape of the letter 'T' reflected in the layout formed by the components VC1, VC2 and L1 in **Fig 1**. The circuit will match the coaxial output of the transceiver to an end-fed aerial or to a coaxial cable feed to the aerial. This design also uses a balun (balance to unbalance) transformer for use with aerials using twin-wire balanced feeder.

This unit will handle up to 5W and operates over the frequency range of 1.8 to 30MHz.

CONSTRUCTION

Inductor L1 is wound on a T-130-2 powdered-iron toroid. The inductor is tapped and fixed to the tags of a 12-way rotary switch. Taps are formed by making a loop about 1cm long in the wire and twisting it tightly. The loops are scraped clean of enamel and tinned with solder ready to be soldered to the tags of the 12-way rotary switch. If the loops are about 1cm long, it is just possible to bend them to fit the tabs of the switch without having to extend them with short wires.

The balun is wound with two wires twisted together, a process known as bifilar winding. These two wires can be twisted together (before winding on to the toroid) by fixing one pair of ends in a vice, the other ends in a small hand-drill.

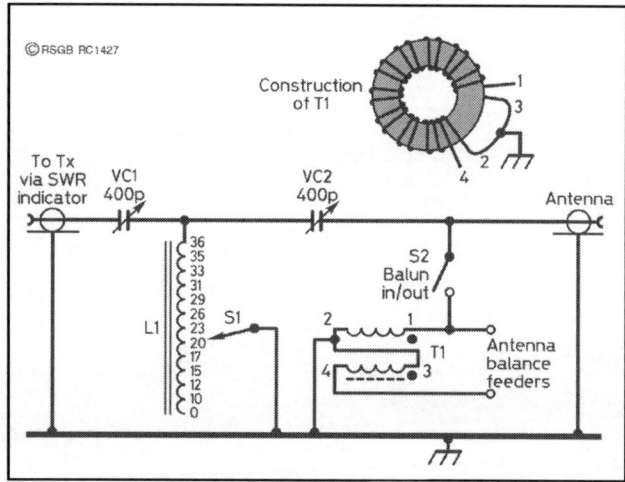

Fig 1. Basic circuit of the ATU.

The drill is then slowly rotated so that the two wires are twisted together neatly. Identify the start and finish of each winding with a buzzer and battery or an ohmmeter. The finish of the first winding is joined to the start of the second, as shown in **Fig 1**.

The two capacitors are twin 200pF polyvaricon capacitors with both gangs connected in parallel to give 400pF max.

You have to drill holes in the box for the control

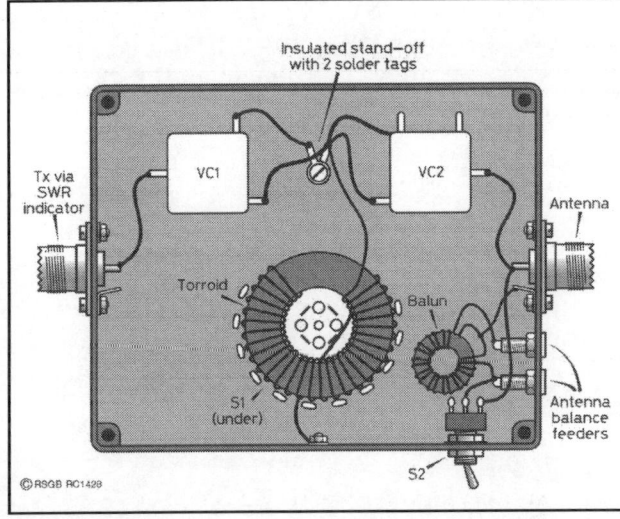

Fig 2. *Internal view of the ATU. A small metal box makes a suitable housing.*

shafts of the capacitors and the switch, and the RF sockets. The switch is fixed using a nut on the control shaft and the two capacitors are fixed using adhesive (hot melt glue is preferred). Take care not to let any tags from the capacitors touch the box as both sides of both capacitors are not earthed.

An appropriate RF socket, such as a SO-239, BNC or phono socket may be used – the choice is yours and should suit your existing equipment. 2 – 4mm sockets may be used for the balanced output. *See* **Fig 2** for the layout, and the photograph for the finished product. It is a good idea to make a graduated dial for each of the three control knobs. An alternative is to use calibrated knobs. You can then calibrate the settings of the three controls so that they can rapidly be reset when you change frequency bands.

OPERATION

The best indication of optimum matching can be

The completed ATU.

achieved using an VSWR bridge; the ATU controls are adjusted sequentially and several times for minimum VSWR. If used for receive only, the best aerial-to-receiver match can be achieved by adjusting the controls for maximum signal.

PARTS LIST

Capacitors

C1, 2 200 × 200pF

Inductors

L1 T-130-2 powdered iron toroid with total 36 turns of 22SWG enamelled copper wire, tapped at 10, 12, 15, 17, 20, 23, 26, 29, 31, 33 and 35 turns from the earth end

T1 12 bifilar turns of 26SWG enameled copper on FT-50-43 toroid

Additional items

S1 1-pole 12-way rotary switch

S2 SPST toggle switch

RF connectors, SO239 sockets or similar (see text)

Aluminium or die-cast metal box

Three plastic knobs for capacitors and switch

Two 4mm sockets for balanced aerial connection

144 To 28MHz Receive Converter   4.1

**A high stability downconverter to let your
HF SDR listen to the 2m band.**

Many of the current Software Defined Radios (SDR) are limited to a maximum tuned frequency of 30MHz. One way to use these SDRs on the VHF, UHF or microwave bands is to use a frequency downconverter, bringing the upper frequencies within the SDR's range

In principle any output frequency up to 30MHz could be chosen for this 'intermediate frequency' (IF), but 28 -30MHz is a common choice.

An older design of converter, such as the Microwave Modules MMC144, could be pressed into service but these tend to suffer in the presence of strong signals due to their high gain. They also tend to exhibit poor frequency stability

Photo 1: Component side of the DDK2010 144MHz to 28MHz converter.

and frequency offset limitations, sometimes being unable to net the local oscillator onto frequency. What is required is a low gain converter, with good frequency and gain stability, together with low noise sidebands. A very low noise figure is not required in this application as the converter is usually preceded by a relatively high gain, low noise transverter (for higher bands) or a masthead preamplifier for 2m. I decided to design and build a new 144MHz to 28MHz receive converter that would meet these requirements. The result is the DDK2010 described in this article.

EXTRA FEATURES

In order to make the converter even more useful, I decided to add the facility for connecting an optional external

high stability local oscillator. If this is derived from a GPS or rubidium source then the receive converter will have exceptional frequency accuracy, as required for some SDR applications. In addition, the internal oscillator is accessible in order to allow it to be used as the LO for an accompanying transmit converter or with a second converter for use in a dual channel synchronous receive system suitable for EME polarisation-dependant reception or simple diversity reception. It is also possible to use this output to feed into an external Reflock unit (see Web-search). The Reflock can then phase-lock the internal 116MHz crystal oscillator. A block diagram showing the various stages of the receive converter is shown in **Figure 1**.

CIRCUIT DESCRIPTION

The DDK2010 receive converter circuit schematic diagram is shown in **Figure 2**.

Local oscillator

A familiar and reliable two stage 'Butler' overtone crystal oscillator is used to generate a 116MHz local oscillator signal that is fed into the LO port of the ADE 1 (MX1) double balanced mixer.

Contrary to normal practice, the oscillator maintaining stage (TR1) uses a switching transistor.

High gain, low noise, transistors, can cause stability problems in this stage. I spent a great deal of time experimenting with different transistors in the Butler two stage overtone crystal oscillator and found that the BSV52 gave consistent and stable operation. Subsequent phase noise measurements (at Microwave Update 2009) confirmed the good performance of this arrangement. The second stage (TR2) is the oscillator limiter, biased for soft limiting so as not to seriously impact phase noise performance. This arrangement was chosen for simplicity over the more usual dual diode limiter. The oscillator output is untuned and delivers about +6dBm to the following mixer stage.

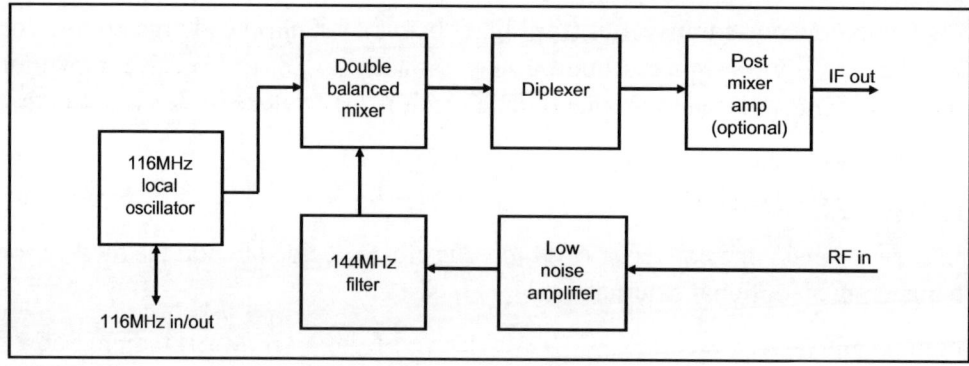

Fig 1: Block diagram of the DDK2010 receive converter.

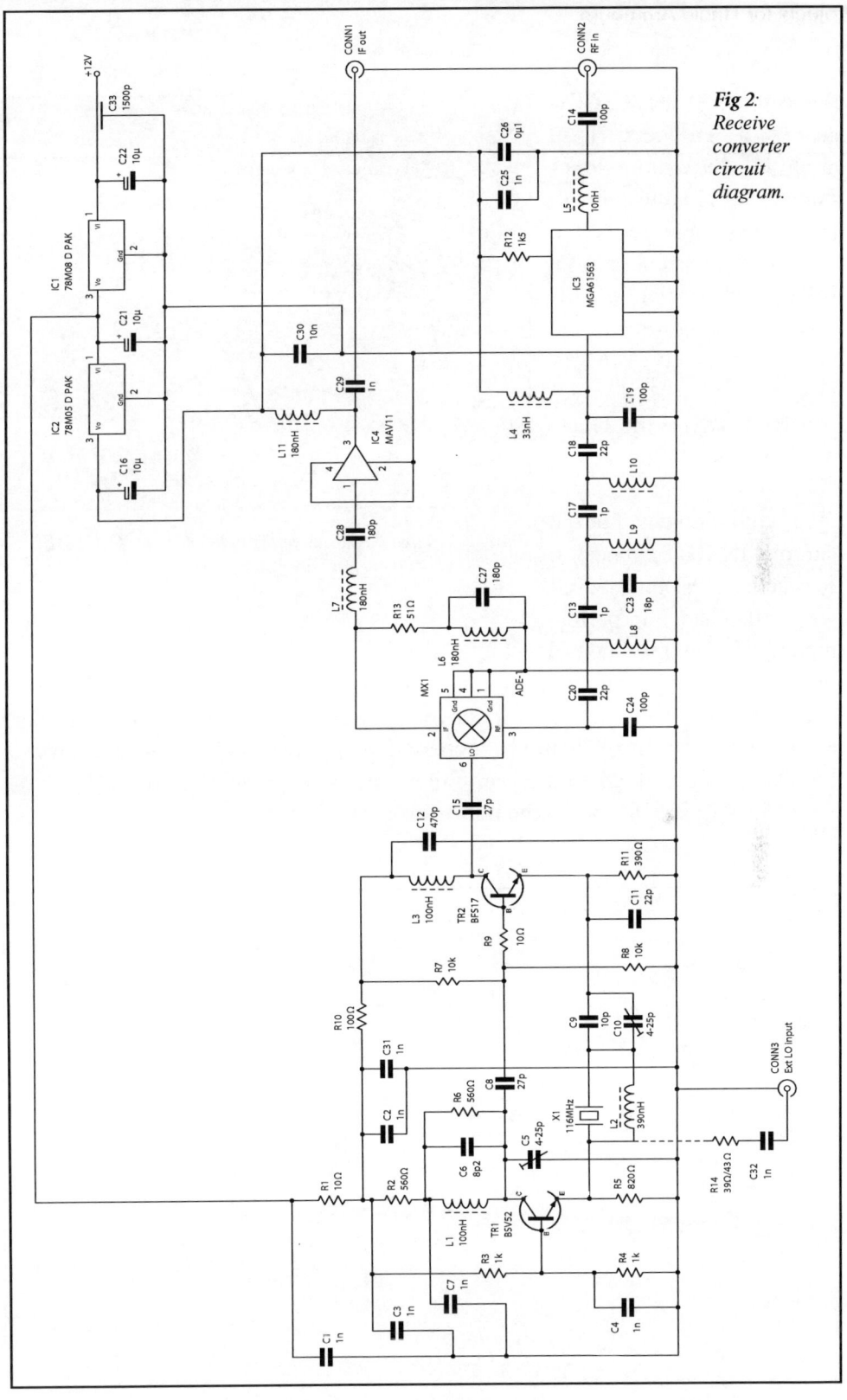

Fig 2: *Receive converter circuit diagram.*

Since the first stage is basically a grounded base amplifier, its emitter input impedance is well defined, as 26/Ie (where Ie is the TR1 emitter current in mA). In normal operation this gives an input impedance of approximately 7Ω. By incorporating a series 43Ω resistor between the external input and the emitter of TR1, an excellent match to 50Ω will be obtained at the external IN/OUT connector. In practice, 39Ω can be used with little change in perfor-

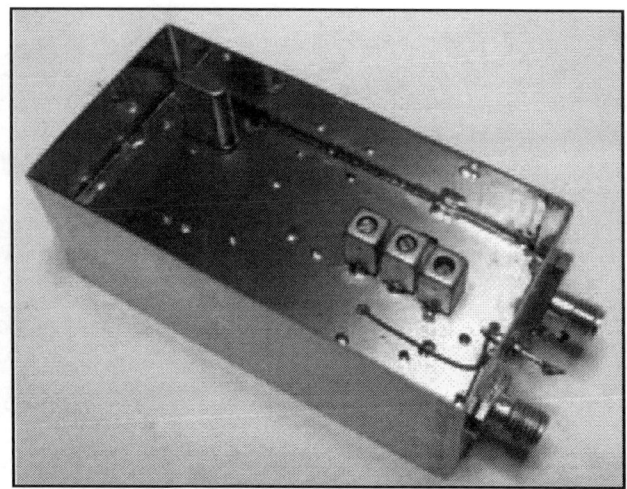

Photo 2: *Ground plane side of the PCB, showing the 116MHz crystal and the band pass filter coils.*

mance. The required input level is -6 to +6dBm. This same connector can alternatively be used as an output, to take a sample of the internal oscillator for use in either an accompanying transmit converter or second receive converter. In both cases the series resistor should be increased in value to minimise interaction with the oscillator, and then a buffer amplifier may be required. A suggested starting value for R14 is 510Ω when the port is used as an output.

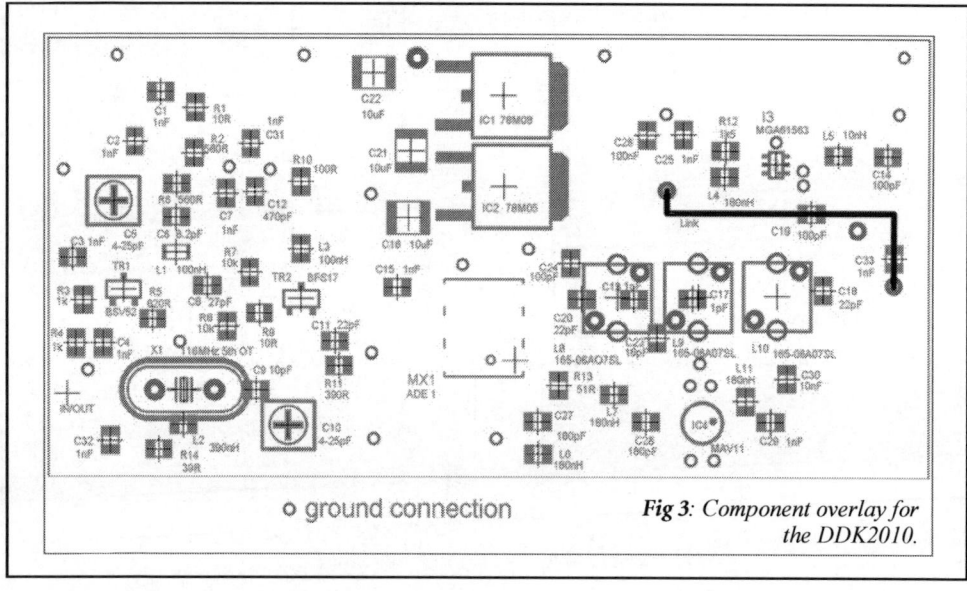

o ground connection

Fig 3: *Component overlay for the DDK2010.*

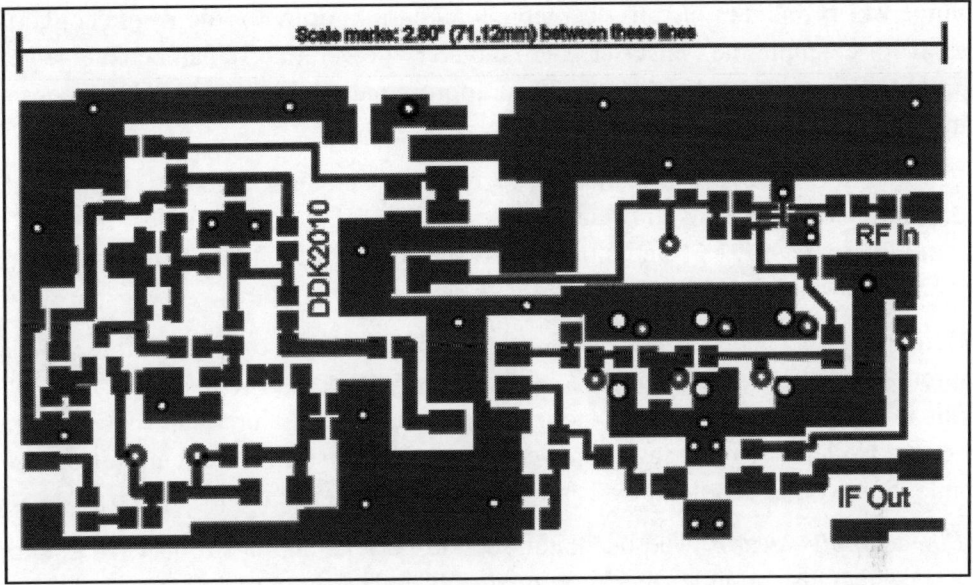

Fig 4: PCB foil pattern for the DDK2010, reproduced not to scale.

Mixer and diplexer. A commercial double balanced diode ring mixer type ADE-1 from Minicircuits Laboratories is used in the DDK2010. Although this is a standard level 7 mixer (ie requiring an LO input at +7dBm), it is slightly under-run in this design because of the low output from the 'Butler' oscillator. If this is a problem then the 8V regulator can be changed to a 10V low dropout device, which will cause TR2 to deliver slightly more output to the mixer.

A 28MHz diplexer consisting of R13, L6, C27, L7 and C28 terminates the mixer IF port in a good 50Ω match at all relevant frequencies. In practice this improves the conversion loss of the mixer and reduces the level of unwanted mixing and injection frequencies appearing at the IF output.

RF and IF amplifier stages. An Avago MGA61563 was chosen for the RF stage because of its good noise figure at 144MHz, high dynamic range and excellent stability. This amplifier provides about 20dB of gain at 144MHz. The MGA61563 is a GaASFET MMIC that is useable from below 100MHz to over 6GHz. Its dynamic range can be altered by the amount of bias current the device is programmed to draw. Resistor R12 sets the bias current to about 45mA from the 5V supply, which provides a high dynamic

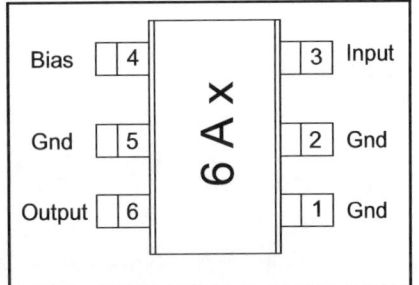

Fig 5: Pinout of the MGA61563. 6A is the device part code and x is the date code.

range whilst maintaining an acceptable total current draw for the whole converter. If for example the converter is to be battery powered, R12 can be changed to 3.9kΩ. This will reduce the current to approximately 22mA, but with a consequent reduction in dynamic range.

IC4 is a MAV11 silicon, broadband, low noise post-mixer amplifier. It is optional, depending on whether you need the extra 12dB of IF gain it provides. If it is not required, leave out L11 and strap across the input and output of the IC4 stage. Another possible reason to omit IC4 is to save a further 40mA.

Band pass filter. The 144MHz inter-stage band pass filter consists of three tuned circuits, top capacitively coupled, with capacitive tapping for 50Ω input and output impedance. The shielded inductors are tuned with aluminium cores. 5mm square Coilcraft 165 series coils were chosen because they are currently available. Some other candidate coils are now obsolescent or on long delivery times.

Powersupply section. The oscillator section gets its supply from 8V regulator IC1, which also supplies the 5V regulator, IC2. The output from the regulator is already well filtered, but it is possible to improve the phase noise performance by a few extra dB by connecting a 100μF aluminium electrolytic capacitor across C21.

The 5V regulator supplies the MGA61563 and the MAV11 post mixer amplifier. As arranged, the MAV11 is slightly current-starved, because it should bias to 5.5V on its output pin. In practice the 5V limitation does not seem to affect the gain or noise figure, although its dynamic range may be slightly reduced. If this is a concern then the MAV11 could be powered directly from the 8V regulator but it will need an additional wire strap from the 8V regulator and a 43Ω current limiting resistor in series with L11. Use a 1206 size resistor due to the high power dissipation.

By using a 'low drop out' (LDO) 8V regulator for IC1, the converter could be powered from a 9V battery if required, since the LDO regulator will function down to about 8.5V input.

CONSTRUCTION

The receive converter uses an etched, 1.6mm thick, FR4, double sided printed circuit board. The mask for this is shown at 1:1 scale in **Figure 4**. The board has a continuous ground plane on the other side. All of the surface mount components go on the track side. Only the crystal, the three adjustable coils and the single insulated wire link are mounted on the ground plane side of the board.

0805 size surface mount parts are used wherever possible, but the pad spacing is such that 0603 size parts could also be used. Low value 0805 size capacitors and

PARTS LIST

Resistors	
R1, 9	10Ω
R2, 6	560Ω
R3, 4	1kΩ
R5	820Ω
R7, 8	10kΩ
R10	100Ω
R11	390Ω
R12	1.5kΩ
R13	51Ω
R14	39Ω or 47Ω

Capacitors

C1, 2, 3, 4, C7, C25, C29, C31, C32	1nF
C5, 10	4 - 25pF Trimmer
C6	8.2pF
C8, C15	27pF
C9	10pF
C11, C18, C20	22pF
C12	470pF
C13, C17	1pF
C14, C19, C24	100pF
C16, C21, C22	10µF, 16V
C23	18pF
C26	100nF
C27, C28	180pF
C30	10nF
C33	1500pF Tusonix feedthrough capacitor

Inductors

L1, L3	100nH
L2	390nH (may not be required)
L4	33nH
L5	10nH
L6, L7, L11	180nH
L8, L9, L10	Coilcraft 165-06A07SL

Semiconductors

IC1	78M08 (500mA)
IC2	78M05
IC3	Avago MGA61563
IC4	MAV11
TR1	BSV52
TR2	BFS17
MX1	ADE-1

Miscellaneous

X1	116MHz 3rd Overtone crystal, HC43/U
Connector 1, 2, 3	SMA 2 hole female bulkhead connectors
Box	Schuberth 34 x 74 x 30mm

fixed inductors, often used in RF designs, are now getting harder to find. Both regulators are rated 500mA and are available in the common DPAK package.

Suitable SMD parts are available in the UK from many suppliers including Farnell, RS and Rapid Electronics. Specialist component part numbers are shown in the parts list. Since there are a large number of alternatives available for many of the other parts it would be impractical to list them all. The adjustable coils are only available in the UK from Coilcraft Europe Ltd. Web addresses for Coilcraft and possible 116MHz crystal suppliers are listed in the Websearch section.

The board is designed to be soldered into a 37 x 74 x 30mm size Schuberth tinplate box to provide screening from external signal pickup. These boxes are obtainable from the supplier listed in Websearch.

The PCB should be cut to 71.6 x 34.5mm in order to be a tight fit into the box. File small notches in the two diagonally opposite corners to clear the overlapping flanges in the box. Do not solder the box together at this stage.

Prepare the PCB by drilling the various through-board holes. The drill size should be no larger than 0.6mm diameter, except for the lugs for the three adjustable coils. These holes should be drilled to 1mm diameter. Solder small-gauge tinned wire through the ground holes. To avoid unwanted short circuits, ensure that the top side (ground plane) copper is cleared around the holes for the pins of the adjustable coils and the crystal as well as

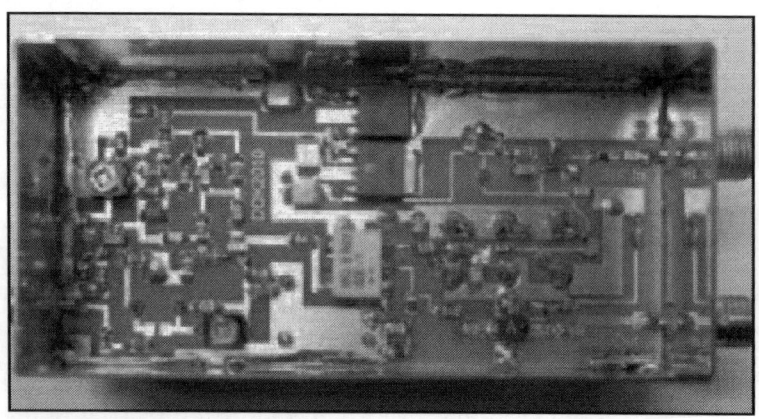

Photo 3: Another view of the 2m converter.

the wire link holes and the regulator DC input. A hand-held 3mm twist drill works well.

Solder all SMD parts onto the PCB using a fine-tipped soldering iron and fine gauge pre-fluxed solder. Start with the resistors and capacitors, followed by the inductors and finally the transistors, integrated circuits and regulators. The regulator 'tabs' should be soldered to the PCB ground plane for heatsinking.

Prepare and solder the insulated single wire link on the ground plane side of the PCB.

Finally, solder the adjustable coils and the crystal to the board, ensuring they are fitted on the ground plane side of the PCB. The crystal should be pushed down snugly onto the ground plane, as should the three coil shielding cans. The coil cans should be soldered to the ground plane along their lower edge.

The most difficult part to mount is the MGA61563,IC3.This device uses aminiature 6-pin RF SOT363 package. Before soldering, ensure that the package is correctly oriented, with the input and output pins in the correct position. *See* **Figure 5** for pinout information. IC3 must be carefully soldered so that its 6 leads are just resting on the 6 PCB pads. It is probably best to initially tack solder one lead, then carefully manoeuvre the package with a pair of tweezers so that the remaining 5 leads lay flat to their respective pads. Take care not to break off any of the 6 leads from the package. Solder quickly, using a small soldering iron and very fine gauge solder – 28 SWG is recommended.

Mark the inside of the box so that the PCB ground plane will be positioned far enough below the rim of the box that the height of the crystal and the adjustable coils can be accommodated without protruding beyond the rim of the box. Also, and most important, mark the correct side of the box so that the corner notches in the PCB are on the right side to engage with the box overlapping flanges. It is easy to get this wrong!

Select and prepare three 2-hole SMA connectors by removing all but 0.5mm of the Teflon™ insulation from the spill of the connector. Mark the position of the three SMA connectors on the inside of the box and drill appropriate size holes (usually 4mm diameter) so that the insulation of the connector just protrudes into the box, but not so far as to interfere with the PCB. The holes for the connectors should be positioned so that the connector spills lay flat onto their respective connector pads. It is advisable to mark and drill holes for the three connectors at this stage as it will be found very difficult to do so after the PCB is mounted in the box, should you change your mind about needing the input/output socket at some later stage. Do not solder the connectors to the box at this stage.

Mark and drill a hole for the supply feedthrough capacitor. If using a Tusonix solder-in feedthrough, you will need a 2mm diameter hole in the box. Do not solder the feedthrough into the box at this stage.

ASSEMBLING THE BOX

In order to ensure the best fit of the PCB into the box use the following procedure. Solder one edge of the PCB into one half of the box by tack-soldering the ground plane side of the board. Ensure the corner notches are on the correct side. Now carefully fit the PCB and its side of the box into the lid of the box. Jig the remaining box side into place inside the lid. The PCB, two sides and the lid should fit comfortably together.

Place the top lid on the assembly and check that it also fits comfortably.

Remove the lower lid. Solder along the overlapping flanges of the box, forming a rigid box section.

Replace the lower lid and remove the top lid. Seam-solder the remaining exposed parts of the box flanges.

Having checked the PCB is level in the box, solder the previously unsoldered side of the PCB to the 2nd box side and then check that the lower lid still fits comfortably in place.

When you are happy that all is well, go back and complete the seam-soldering of the PCB to the inside of the box. This will need to be soldered on both the ground plane side of the PCB and along some of the track side.

Solder the connectors and the feedthrough capacitor on the outside of the box, making sure that you include a 2mm hole diameter solder tag under the feedthrough before soldering it in place. The lack of a suitable supply grounding point can be an annoyance when testing!

Solder the connector spills to their respective PCB pads.

Connect a short insulated wire from the feedthrough capacitor to the regulator input.

Photos 1 and 2 show the top and bottom of the completed converter.

ALIGNMENT

Before connecting any power, check with an ohmmeter that there is no short between the supply input feedthrough capacitor and ground. If all is well, connect a +12V current limited power supply set to less than 150mA to the feedthrough capacitor and 0 volts to ground. Check the total current taken, which should be no more than about 100mA, depending on whether IC4 has been fitted. If it is higher, check for faults such as short circuits, incorrectly placed components or reversed polarity tantalum capacitors. If all is well then proceed to set up the crystal oscillator. This is easily done by listening to the output from a receiver tuned to 116MHz. Most scanners and aircraft band receivers will cover this frequency. Use FM or AM to listen for a distinct change of noise from the receiver when C10 is adjusted. C5 will need to be set near minimum capacitance to allow the oscillator to operate at 116MHz. If the receiver is within a few metres of the converter the noise change will be very apparent. The final local oscillator frequency can be set more accurately later. The Butler oscillator is a very reliable starter. This version, with the BSV52 first stage, is particularly docile. Even very lazy crystals should work well in this circuit.

If you have access to a spectrum analyser and signal generator, then you will probably already know how to set up the adjustable coils of the band pass filter by adjusting for maximum IF output at 28MHz with the signal generator set to 144MHz. Do not set the signal generator to greater than -20dBm output, to avoid saturating the converter output.

If you do not have access to test equipment then connect a HF receiver, tuned to the 28MHz band, to the IF output. Connect a 144MHz antenna to the RF input. Tune the receiver to a strong local repeater output channel or to a local beacon of known frequency. Assuming a signal can be heard, adjust L8, L9 and L10 for maximum signal. If no signal can be heard, you should still hear a small but noticeable increase in noise output as the 144MHz bandpass filter is adjusted to cover the correct frequency range. The limited adjustment range of the coils in the filter should prevent any chance of mistakenly tuning to the image at 88MHz. When correctly adjusted, the band pass filter will exhibit a nearly flat frequency response across the 144 to 146MHz frequency range.

If you can now hear a beacon or repeater, but it is slightly off frequency, adjust

C10 to obtain zero beat when using the receiver on USB (or LSB) mode. It may be necessary to slightly adjust C5 to bring C10 within range. In the event that you cannot move the frequency of the oscillator onto exactly 116.000MHz it may be necessary to connect the optional inductor L2 across the crystal. This is probably not required, but may help. If you are not within range of a repeater or beacon use another, suitably accurate, signal source to adjust the local oscillator frequency.

There are no other adjustments to make and the converter is now ready for use.

PERFORMANCE

The measured performance of the prototype converter is shown in **Table 1**. Note that the alignment of the bandpass filter, whilst not critical, must be done carefully as small changes in filter loss, due to poor alignment, can result in an unacceptably high noise figure in such a low gain converter design. The overall noise figure of the converter is lower (better) without the MAV11 stage. This may not sound very intuitive but is due to the 'second stage' noise contribution of the MAV11 stage. The same degradation in noise

figure will be experienced when the converter is fed into a typical 28MHz super-het receiver or SDR receiver.

EXTENDING THE USE OF THE RECEIVE CONVERTER

The provision of an external local oscillator input/output makes the DDK2010 very flexible since it is possible to connect an external, high stability 116MHz source in place of the internal crystal oscillator. If this is GPS-locked then the receive converter can become part of a completely frequency-locked receiver/transverter system for any of the bands on or above 144MHz. Alternatively, the internal 116MHz signal can be extracted from the input/output socket and used to feed into an accompanying transmit converter to produce a high performance transverter for 144MHz.

For 144MHz moon bounce (EME) enthusiasts, the 116MHz output can be fed into the corresponding input/output socket of a second DDK2010 receive converter (keep R14 as 39Ω or 43Ω) to produce a dual RF input system for use with a polarity sensing receiver that uses one converter on the horizontal antenna array and a second converter on the vertical one.

It is also possible to replace the MAV11 post mixer amplifier with a higher gain device such as the MAR6. The receive converter will have a similar noise figure with a gain up to 8dB higher, but the dynamic range will be reduced. An additional 91Ω resistor will be required in series with L11 to obtain the correct bias for the MAR6.

PROTOTYPE RECEIVE CONVERTER PERFORMANCE

Noise figure 2.5-3dB Gain 17.5dB Input 1dB

compression point -16dBm Bandwidth 3.0MHz Image response Greater than -80dB Current consumption 100mA at 12V

Sam Jewell, G4DDK
sam@g4ddk.com
March 2010

WEBSEARCH

Reflock: *http://gref.cfn.ist.utl.pt/cupido/reflock.html*

SMD parts are available from many suppliers including *www.farnell.com*, *http:// uk.rs-online.com/web* and *www.rapidonline.com*

The Avago MGA 61563 data sheet can be downloaded from *www.avagotech.com/ docs/AV02-1471EN*

The adjustable coils can be obtained from *www.coilcraft.com/general/sales_ eu.cfm* 116MHz crystals can be sourced from *www.quartslab.com* or *www. eisch-electronic.com* Schuberth Tinplate Boxes are available from *www.alan.me-lia.btinternet.co.uk*

Speech Processing

There is nothing like a demonstration to illustrate a point, so I would strongly recommend this experiment. On the FM waveband, tune in BBC Radio 4 and adjust the volume to give a comfortable listening level on ordinary speech. Then, retune to a commercial 'pop' station. Whether the content is music or speech, you will probably reach for the control to turn down the volume. The reason for this is the subject of this article, although it is its application to amateur radio that concerns us most. Just like us (but perhaps even more so), the broadcasting companies must not over-modulate or over-deviate, so why is speech on one broadcast so much louder than on another?

FIRST STEPS

The waveform (as might be shown on a cathode ray oscilloscope) of male speech, taken from a discussion programme on BBC Radio 4, is shown in **Fig** 1(a). **Fig** 1(b) shows the waveform of male speech from a commercial radio station, and **Fig** 1(c) the waveform of music taken from the same station. All three recordings were made with the same audio gain setting (with no audio AGC), and are of the same length, so they are thus directly comparable. The equipment used was an Icom IC-PCR1000 receiver, a Soundblaster® PC sound card, and proprietary software for recording wave files. To analyse these

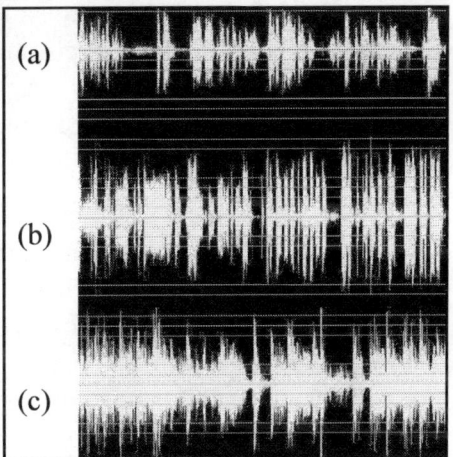

Fig 1. (a) A section of male speech from BBC Radio 4. (b) A section of male speech from a commercial radio station. (c) Vocal pop music from the same commercial radio station.

waveforms quickly, I wrote a short program in Visual Basic 6.0, which analyses the waveform and calculates the power (in arbitrary units) in each waveform. The results are summarised thus:

Radio 4 speech:	Power = 138
Commercial radio speech:	Power = 504
Commercial radio music:	Power = 522

For these random selections of programme output, you can see that there is almost four times as much power in commercial radio speech as there is in Radio 4 speech, and that the music played on commercial radio is only marginally louder than the DJ's pearls of wisdom! This process of 'making everything louder' is called dynamic range compression, contrast compression or, simply, compression. As we amateurs are concerned only with speech, we know it as speech compression or speech processing and, for reasons to be discussed later, we use it only when working on SSB.

THE PROBLEM

Speech is remarkably 'spiky' in nature, as Figs 1(a) and 1(b) show when viewed on a cathode-ray oscilloscope. This spikiness causes us problems when we have to set the modulation level on our transmitters. On sideband transmissions, over-modulation causes splatter at the very least, and on FM transmissions it produces over-deviation. Both of these must be avoided at all costs.

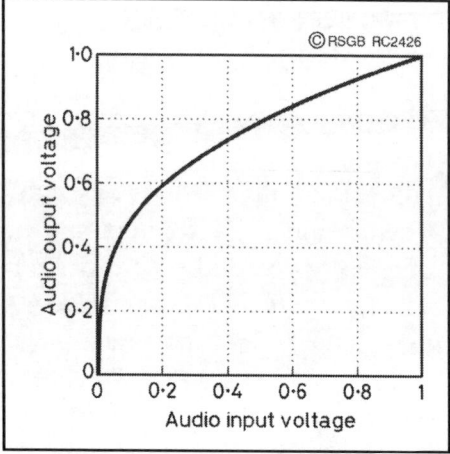

Fig 2. Using a non-linear transfer characteristic as the basis of a speech processor.

To do this, we must set the transmitter drive so that the peaks of our audio waveform do not over-modulate. To illustrate this, use **Fig 1**(a), and lay a ruler on it horizontally so that it just touches the biggest negative-going peak (we choose a negative-going rather than a positive-going peak, so that the ruler does not hide the waveform). You will now notice one thing – that the average modulating voltage (compared with our self-imposed maximum) is very low. In practical terms, our transmission will lack 'punch', and will not be heard very well under difficult conditions.

SOME SOLUTIONS

In very basic terms, what we need to do is to turn up the audio gain when the speech is quiet and turn it down when the speech is loud. Unfortunately, these variations of loudness occur very quickly, often between syllables, and the technique is something which cannot be done manually.

Two basic methods are used, one of which involves using an amplifier of non-linear transfer characteristic; the other uses automatic control of the audio gain. It should be understood at the outset that, whatever technique is used, it distorts the speech waveform. This distortion is (hopefully) controlled so as to improve the intelligibility, even if the result is sometimes unnatural.

Using a non-linear transfer characteristic is, in theory, probably the most attractive solution. A transfer characteristic is the graphical representation of the output

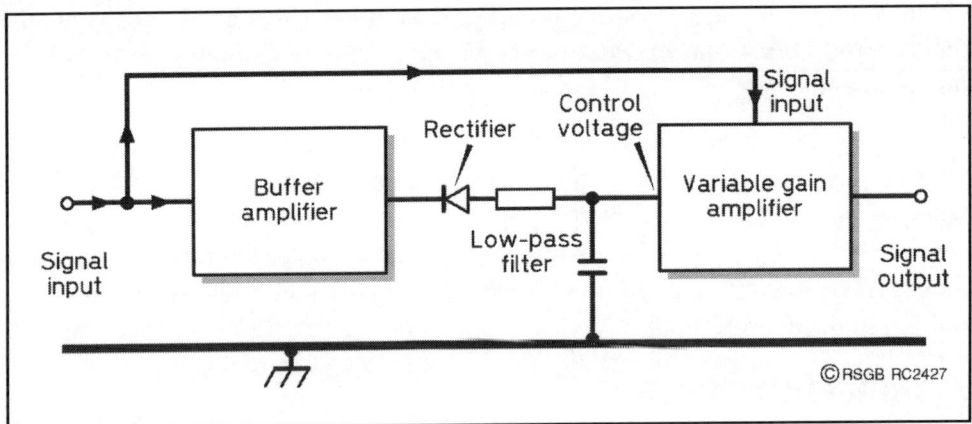

Fig 3. Block diagram of a simple audio AGC circuit.

voltage from a device compared with the input voltage. Normally this should be a straight line, indicating that the output and input voltages are directly proportional.

To achieve the desired compression, look at the characteristic of **Fig 2**; notice that, for example, a small input of 0.1V will produce an output of about 0.5V (a gain of about 5), 0.2V will produce an output of about 0.6V (a gain of about 3), whereas a large input of 1V will produce an output of 1V (a gain of unity). Provided the input voltage never exceeds 1V, the circuit will produce an increasing gain at lower voltages. The main problem with a technique such as this is that a circuit to implement it is quite difficult to design. It does have the advantage (see later) that nothing physically changes as a result of changing input levels.

The other techniques are simpler to implement, but have their own disadvantages. Manually altering the audio gain control may be completely impractical, but circuits can be made which will achieve the same result automatically.

The operations involved are straightforward:

(a) Rectify the audio voltage. The average value of an audio signal is zero, so rectification is needed to 'lop off' the lower half of the signal so that it does have a mean value.

(b) Feed the rectified signal into a low-pass filter, a simple connection of one resistor and one capacitor (see **Fig 3**). The output from this filter is the mean value of the signal over short time periods, the audio components having been short-circuited by the capacitor.

(c) This varying mean value is then used to control the gain of another audio amplifier, the gain being reduced as the mean value increases. A block diagram of the circuit is shown in **Fig 3**. A buffer amplifier is a device which helps to separate the rectifier and filter from preceding circuits and to supply enough gain to operate the rectifier cleanly.

A simple gain control circuit, eminently suitable as the basis for experimental work, is shown in **Fig 4**. An n-channel JFET is used as one of a pair of feedback resistors in a standard non-inverting op-amp circuit. Connected this way, the resistance between the JFET drain and source depends upon the voltage applied to its gate. The overall gain [1] is adjustable between 1 and 1000 by changing the voltage at the control input.

ATTACK AND DECAY

While **Fig 4** is quite a simple and acceptable circuit to use for automatic audio gain control, the derivation of the control voltage to operate it can be quite tricky. The simplest circuit, comprising the rectifier, capacitor and resistor shown in **Fig 3**, is a good starting point. The values of the resistor and capacitor need to be varied in order that the circuit will act quickly to reduce the gain when a loud sound suddenly appears (this is known as 'attack'), yet will take a little longer to return to its initial gain (known as 'decay') in case another loud sound follows on quickly. Getting the attack and decay right is quite an art, and a circuit more complex than that of **Fig 3** is usually needed.

THE PROBLEMS

No circuit like this is without its problems. Correct attack and decay are obvious candidates, and the solution is somewhat subjective, because no two people speak in exactly the same way. However, the overriding problem with all AGC-based speech processing circuits is that a loud sound must get through to the output before the circuit can react to reduce the gain. This means that, in unfortunate cases, transient peaks which are louder than they ought to be will get through to the output and could still cause transmitter over-modulation.

The overall drive must be backed off a little to allow for this. Nevertheless, speech processing, when properly used, is of great benefit to the SSB station working in crowded band conditions.

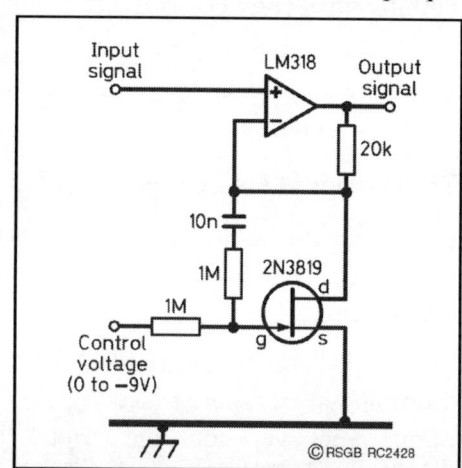

Fig 4. A simple amplifier circuit where the gain is dependent upon the control voltage. The circuit supply voltage should be around ±15V.

In professional circuits using this principle, the signal between the main input and the input to the gain-controlled amplifier in **Fig 3** is subject to a short delay; which means that the gain can be turned down just in time for the delayed signal which caused it to reach the amplifier!

OTHER METHODS

Speech processing may also be achieved (and more effectively so) by performing the control operation on the RF signal, or by using DSP. Both these techniques are well beyond the scope of this article, however.

FINALLY

We have not discussed speech processing in the context of FM. FM is essentially a short-distance mode, and interference from adjacent stations is not a problem, thus removing the need for processing.

If circuits are useful, you can bet that they have been made into integrated circuits! The speech processor is no exception, and one such device is the VOGAD in an 8-pin DIL chip, the SL6270 (Maplin order code UM73Q).

REFERENCE

[1]   The Art of Electronics, Horowitz and Hill, Cambridge University Press, 1988, p240/41.

An Audio-Driven S-Meter For DC Receivers

Direct conversion (DC) receivers do not normally have automatic gain control (AGC). The reason for this is primarily that they do not have intermediate frequency (IF) amplifiers, whose gain can be varied to control their output without introducing too much distortion. If the amplifier's output is to be held reasonably constant, then the AGC voltage must track the received signal strength and is therefore a convenient means of driving a signal strength meter.

Alas, the DC receiver does not have this feature. The mixer produces an audio output and is normally directly followed by a low noise pre-amp of fixed gain and wide dynamic range. The primary gain control of the receiver is usually placed after the audio pre-amplifier. With such an arrangement, the audio signal across the gain control is directly proportional to the incoming RF signal over the linear range of the receiver's front end, so why not use the audio output to drive a signal strength meter via a rectifying circuit? The answer to this question is perhaps best appreciated by considering the actual signal strengths typical of the HF bands and how they vary.

S-UNITS

Amateur signal strength reports use the scale 1 to 9, S9 being a noise-flattening solid signal. Way back in the 1940s, commercial receiver manufacturers tried to standardise on a value of 50µV RMS (into a 50Ω load) as being an S9 signal and this is still a good figure to use. When listening on a receiver with an S-meter calibrated to this standard and not having excessively narrow passband, the sound of the signals seem to match the indication, so let's use 50µV as the S9 level and consider what lower levels mean.

One S-point is taken to mean a 6dB change in signal strength, which is a clearly discernible level change to the ear. If the signal is getting weaker, then one S-point down is -6dB. A -6dB change occurs when the signal strength is halved in value, ie reducing from 50µV to 25µV is a one S-point drop. If you continue to halve the signal level eight times you arrive at 0.19µV as equivalent to S1, which is clearly a very weak signal. In fact, constructing an HF receiver which will resolve an S1 signal is no mean achievement.

Having established the signal range, if you attempt to display the signal strength using a conventional, linear, op-amp style audio rectifier, there is a problem. For example, if you use an ordinary 0 – 10 scaled meter as an indicator with scale digit 9 being S9, then an S8 signal will appear at 4.5 on the scale and S1 would be difficult to read at all. Alternatively, re-scaling the meter to indicate S-points results in a very cramped scale.

Clearly, for an S-meter to be easily read it must display 6dB steps as equal increments. In other words it must have a logarithmic response. This is very difficult to achieve directly in a moving coil meter and is most easily accomplished using non-linear driving circuits. There are several ICs on the market with logarithmic responses, but one of the easiest to use (and the cheapest) is the old Motorola MC3340P. This is a straightforward voltage-controlled attenuator chip and it forms the heart of the following circuit.

USING THE MC3340

The MC3340 is a wideband attenuator chip, claimed by the manufacturers to have a staggering 80dB range. The control characteristic is substantially linear, in dB/V, provided you do not operate at gain levels of 0dB and above. The control voltage range is a little difficult to handle, as it is somewhat dependent upon the supply voltage and has a positive offset of about 2.6V at maximum gain. The manufacturers' performance curves, which I obtained from a data CD ROM issued by Farnell [1] were helpful in establishing the general shape of the IC's response, but the optimum working point was found, in time-honoured fashion, by painstaking measurement. Motorola's ICs do, however, seem to maintain their characteristics from sample to sample; the two I tried were almost identical in performance, even though they were purchased some 15 years apart!

Fig 1 is a block diagram of the meter driver. It is designed to take audio from across the AF gain control and the S9 level is assumed to be 400mV RMS. The input resistance of the attenuator chip, IC1, is about 20kΩ and may need to be buff-

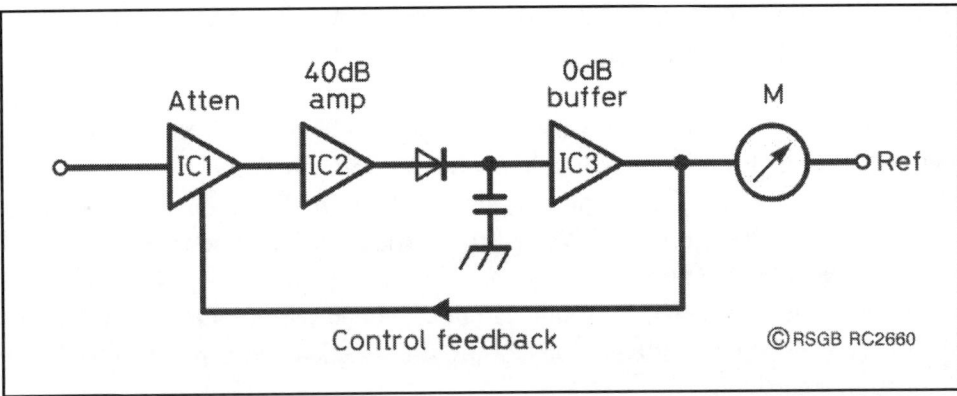

Fig 1. Block diagram of the audio-driven S-meter..

ered if the AF signal comes from a source resistance much above 5kΩ. IC1 is supported by an AF amplifier, IC2, a diode peak rectifier, and a buffer stage, IC3. It was found that for best linearity, IC1 must be operated in a region where it introduces a significant loss of signal and the 40dB amplifier stage IC2 is included in the loop to recover the audio to a level suitable for rectification. After peak rectification and buffering, the resulting DC signal is returned to IC1's control pin, forming a negative feedback loop. This feedback has a twofold effect. Firstly, it stabilises the output of IC1 and secondly, it helps to linearise the control characteristic.

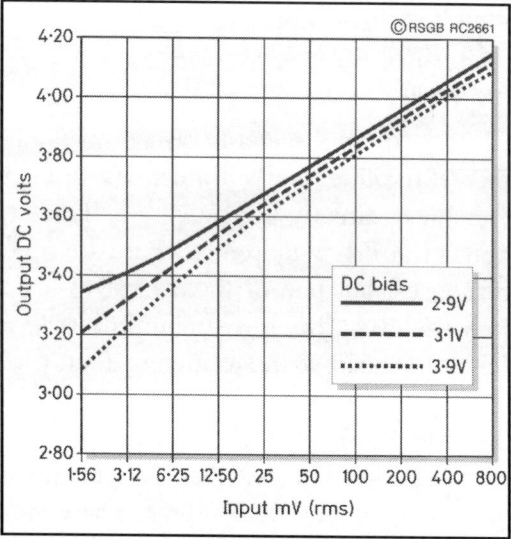

Fig 2. MC3340 output for various bias voltages.

Fig 2 shows the measured response of the circuit, based upon an S9 audio level of 400mV RMS. The three traces illustrate the effect of changing the zero signal bias voltage. Best overall linearity is achieved with a zero signal bias set at around 3.15V when using a 12V supply. Note the equal increments in control voltage for each doubling of the signal strength.

Thus, if the indicating meter has its negative pole set at 3.15V, its response can be scaled using a series resistor to show S-units directly. A linear increment range of ten S-points can be displayed with this arrangement, which makes the circuit well suited to drive a standard 0 – 10 scaled meter. Again, scale 9 is set as S9 and is achieved with an audio signal input of 400mV RMS. However, this limits the maximum indicated signal to one S-point over S9. Compared to a commercially-produced receiver, with an S-meter scaled to 60dB over S9, this performance may look limited until you realise that such an indication means a signal 1000 times more powerful than S9 and is perhaps of limited application! Practically speaking, since this circuit is intended to be added to DC-type receivers, which usually employ an RF attenuator to cope with extra-large signals, if steps that are multiples of 6dB are used, the range can be extended without the need to modify the meter scale. An RF attenuation of 6dB simply adds one S-point to the meter indication, and so on.

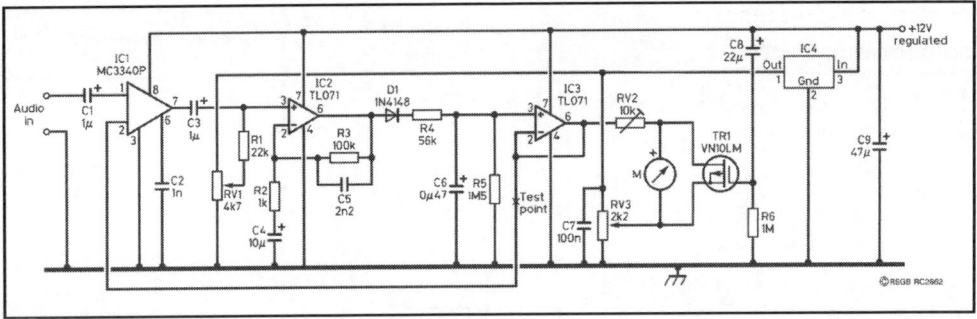

Fig 3. Complete circuit of the audio-driven S-meter.

GENERAL CIRCUIT POINTS

The final circuit is shown in **Fig 3** and PCB layout in **Fig 4**. Stable DC supplies are essential. A 12V regulated supply was available in my DC receiver and I used that in conjunction with a low power regulator, IC4 (78L05), to fix the 5V bias rail voltage. The circuit will work just as well from a 15V regulated supply, but IC1's bias voltage would need to be adjusted to 4V. Operation from lower supply voltages is not recommended, as linearity suffers.

The meter used should have a full scale sensitivity in the range of 100 to 400µA.

Small meters with light pointers are the best types to use as they can follow variations in signal strength quickly, the movement having low inertia. Good VU indicators meet this requirement but are usually scaled in 3dB steps and need rescaling. Helpfully, because the meter is driven from the output of an op-amp, the actual full scale sensitivity is of little importance - provided you do not exceed the maximum current output of the amplifier. The value of RV2 may

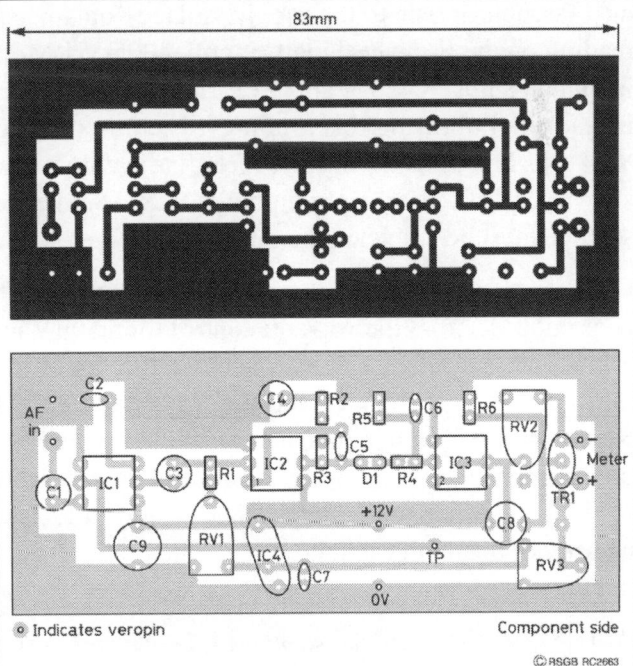

Fig 4. PCB layout and component placement.

be varied to suit the meter if out-side the range specified.

To overcome switch-on tran-sients, which tend to cause vig-orous full-scale pointer move-ment, C8 is used to bias the MOSFET, TR1, temporarily into conduction and short out the meter. After a second or two, when the circuit capacitors have reached their working voltages, C8 has also charged via R6 and the circuit reverts to its normal state as TR1 switches off.

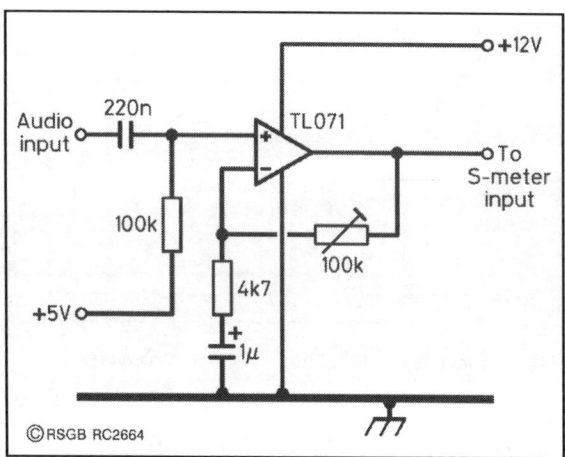

Fig 5. Circuit of the optional pre-amp.

INITIAL ADJUSTMENT

Setting up the circuit is quite straightforward, but does need a DC meter and an audio signal generator. Firstly, short the audio input to guarantee no signal. Adjust RV2 and RV3 to mid-range. Connect the DC meter to the test point or the output of IC3 (pin 6). Now switch on the supply and adjust RV1 for the correct zero sig-nal bias, appropriate to the supply voltage you are using. Next, set RV3 for a zero reading on the S-meter. Finally, apply 400mV RMS at about 800Hz to the audio input and adjust RV2 for an S9 reading. Repeat this cycle of events until stability is reached. Following this routine, it is comforting to reduce the AF signal input to half its level and see that the meter indication reduces by one S-point. It is pos-sible to repeat this action right down to S1, but beware, you need a well earthed/screened test bench when the input signal level is down to a few millivolts.

Connecting the meter to the receiver audio requires that an audio level of 400mV be produced across the AF gain control for a 50μV input signal. This may require a low gain amplifier, either to achieve the audio level or to prevent loading of the audio stage in the receiver. **Fig 5** is suggested if such an amplifier is needed. Remember, the input resistance of the MC3340P is about 20kΩ. Unless this is high compared to the output resistance of the receiver audio pre-amp stage, it may load the amplifier and perhaps cause distortion. Also, if this pre-amp is used, the polarity of the input capacitor C1 should be reversed. This is because the pre-amp output is at +5V.

At G3DXZ, the current low-band DC receiver sports an edgewise S-meter, ob-tained surplus and re-scaled. With many of this type of meter, the scale is a strip of paper that simply clips into the plastic housing and is very easy to remove, lay flat and re-scale or simply replace. Rescaling also has the advantage of allowing you to shift the 10 S-point range of the circuit to start, say, at S3 and give some over-9 indication if that is preferred. The real bonus in using one of these surplus indicators is that they often have a built in lamp to illuminate the scale, and that does add a touch of class!

PARTS LIST

Resistors (all fixed resistors, 0.25W 5%)
R1 22k
R2 1k
R3 100k
R4 56k
R5 1.5M
R6 1M
RV1 4.7k mini carbon pre-set
RV2 10k mini carbon pre-set
RV3 2.2k mini carbon pre-set

Capacitors
C1 1µF 63V electrolytic
C2 1nF 63V mylar
C3 1µF 63V electrolytic
C4 10µF 63V electrolytic
C5 2.2nF 63V mylar
C6 0.47µF 63V electrolytic
C7 100nF 63V mylar
C8 22µF 16V electrolytic
C9 47µF 16V electrolytic

Semiconductors
IC1 MC3340P
IC2 TL071
IC3 TL071
IC4 78L05
TR1 VN10LM
D1 1N4148

Miscellaneous
M1 100 – 400µA

REFERENCE

[1]Farnell, Canal Road, Leeds LS12 2TU. Tel: 0113 263 6311.

This transceiver was developed as a project that would hopefully appeal to the relatively inexperienced constructor, although it is probably not ideal for the complete beginner. The ability to handle a soldering iron and identify components is necessary, as is the ability to read and work from a circuit diagram, but a lot of detail is included to make construction as straightforward as possible.

DESIGN PHILOSOPHY

The aim was to produce a basic transceiver with minimum component count that required little in the way of tools or test gear to get working, but which would give creditable performance.

A tall order, perhaps. The first step was to decide exactly what was required. The obvious choice was a single-band QRP rig. On the receiver side, I opted for direct conversion, as a superhet design would be far more complex. For the same reason, I decided on CW rather than SSB.

Now, if you are going to operate QRP CW for the first time, the ideal place to do it (in my opinion) is on 80m. On this band, a watt or two will give you QSOs all over the UK and well into Europe, a fact that I discovered around 30 years ago, before QRP operation really took off. I reduced the output of an old valve transmitter to about 3W, and to my surprise found that it seemed just as easy to make QSOs with this as it did with my 100W transmitter.

There are many circuits for this kind of transceiver around. A lot of them seem to

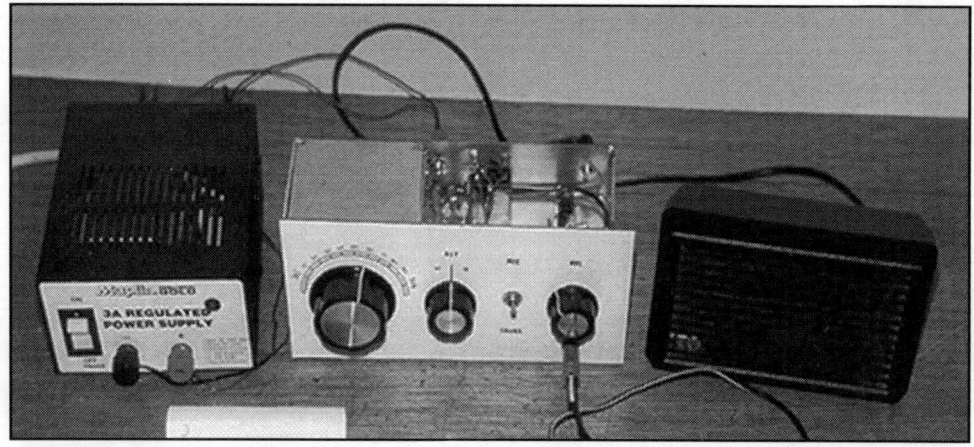

The completed transceiver (centre)

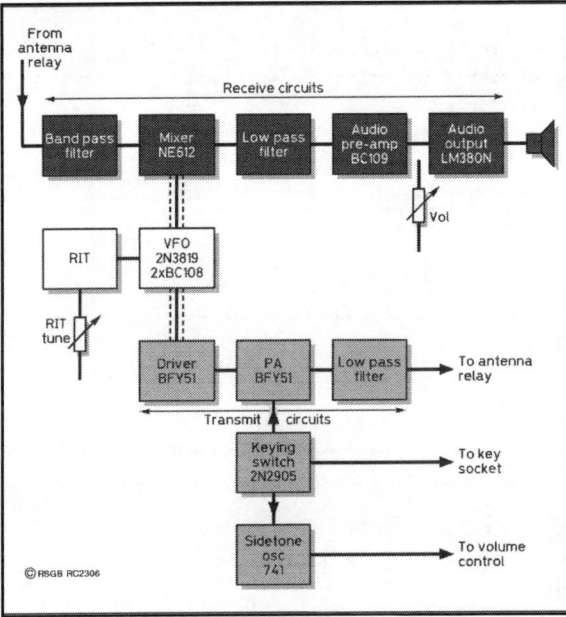

Fig 1. Block diagram of the 80m CW transceiver. Power supplies have been omitted for clarity

be severely limited by over-simplification. I consider that VFO (variable frequency oscillator) control is highly desirable, but that RIT (receiver incremental tuning) is absolutely essential. If you are using crystal-control for transmit and receive, you have to rely on the other station being slightly off your frequency in order to produce the necessary beat note, or be able to shift the frequency of your oscillator slightly. With RIT and a VFO, it is a simple matter to 'net' onto another signal, and then adjust your RIT for a comfortable note. Another highly desirable feature is a sidetone oscillator. This enables you to hear what you are sending, and does not add greatly to the overall complexity.

The method of construction is at least as important as the electronics in a practical project.

I opted to use double-sided copper-clad board to make a chassis / panel arrangement, with a rear apron and a screened enclosure for the VFO. Cutting and drilling are kept to a minimum, and the need to manufacture printed circuit boards avoided by mounting most components (with the exception of those in the VFO, which are mounted on a small piece of matrix board) in a 'dead bug' fashion. More on that later.

DESCRIPTION

The block diagram of the transceiver is shown in **Fig 1**. The VFO is common to transmit and receive.

On receive, the incoming signal is passed from the aerial socket by the change-over relay to the band-pass filter (which greatly attenuates out-of-band signals) and then to the mixer where it is mixed with the VFO signal to produce an audio beat note. Most operators favour a note of 800 – 1000Hz, so the VFO will be offset by this amount. Because there will be many in-band but unwanted signals at the mixer input, there will be many more at the output after the mixing process.

Although most of these will be beyond the range of hearing, some may be strong enough to overload the audio stages.

The low-pass filter will all but remove those at radio frequencies, and also attenuate the higher audio frequencies. The audio preamp now amplifies the wanted signal to a sufficient level to drive the output stage. On transmit, the VFO signal is amplified by the driver and then the PA to give an output in excess of 1W. Harmonics are attenuated by the low-pass filter and the signal is passed to the aerial

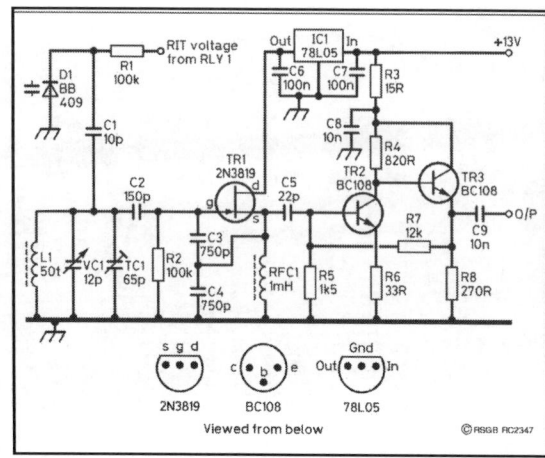

Fig 2. The VFO consists of a buffered Colpitts oscillator.

socket by a change-over relay. Keying is by means of a 'keying switch' (electronic, not mechanical) which keys the 13V supply to the PA. This keyed 13V also supplies the sidetone oscillator, the output of which passes to the audio output stage via the sidetone level control.

The RIT in this design allows the operator to tune approximately ±1.5kHz of the frequency on receive. The position of the RIT control does not, of course, affect the transmit frequency.

THE VFO

There is nothing remarkable about the VFO (**Fig 2**). Similar circuits are to be found in many pieces of equipment. The oscillator transistor is a 2N3819 field-effect transistor (FET). This operates at a low power level so the generation of excess heat, the enemy of VFO stability, is minimised. The supply is stabilised at +5V by IC1. The buffer stage consists of two BC108s, chosen because they are cheap, common, and I have a bag full of them. Most amateurs must have a few lying around. The coil L1 is wound on a T50-2 toroidal core. Wound with the number of turns specified, you should have a VFO that works on the correct frequency, which might not be the case if any old former and core were used (I don't know of any current source of new coil formers). In practice, frequency stability is quite adequate.

Mention must be made of the tuning capacitor, VC1. The tuning range is determined by the value of this component. 12pF allows coverage of the whole of the CW segment with a little to spare at each end, while 10pF will just cov-

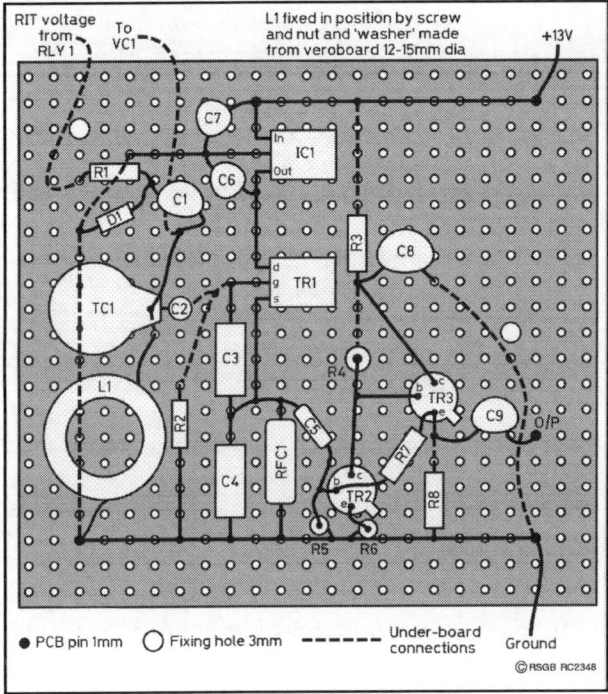

Fig 3. The VFO is built on matrix board.

er the required range. As I prefer to have a little overlap, I used 12pF. With the amount of bandspread that this provides, the tuning rate is comfortable without a slow-motion drive. This, of course, simplifies mechanical construction. The value of a variable capacitor can be reduced by removing some of the plates. Actually, it is the number of gaps that determines the value, so if you divide this number into the value, you will know how many picofarads each gap contributes. The excess plates should be carefully removed with long-nosed pliers, taking care not to damage the remaining ones or over-stress the bearings or shaft. So, if you have a component with a value that is a little too high, you can still use it, although I wouldn't recommend trying to reduce the value by more than 50% as the results become less predictable.

It was decided not to build the VFO 'dead bug' fashion for two reasons: (a) the rigidity (very important in a VFO) wouldn't be as good as it could be; and (b) thermal stability. With the frequency-determining components in contact with the chassis, they would be more prone to sudden changes in temperature than if they were mounted independently. As printed circuits had been ruled out, the VFO was built on a 21/4in × 21/2in piece of 0.1in matrix board. This, in turn, was mounted using two 12mm M2.5 screws, with extra nuts as spacers to keep the board clear of the chassis. *See* **Fig 3** for the layout of the VFO board.

It turned out to be essential that the VFO be fully enclosed, not for the obvious reason of RF screening but because the slightest draught caused unacceptable drift. Holes are drilled in the screens to allow connections to pass through. Obviously this should be done before assembly. Initially the lid can be left off. Decoupling capacitor C40 should be mounted close to where the lead carrying the RIT voltage enters the enclosure.

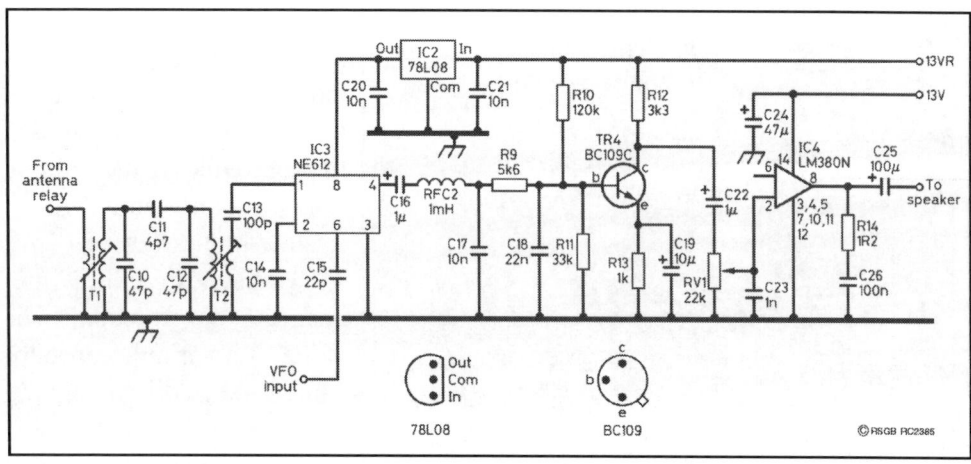

Fig 4. Circuit of the receiver. The VFO input is via a short length of miniature coax from C9

THE RECEIVER

Referring to **Fig 4** and the detail in **Fig 5**, the band-pass filter comprises T1, T2, C10, C11 and C12. The transformers are from Toko, which make circuits such as this easily reproducible.

Next comes the mixer. The NE612 is a useful device, containing an oscillator as well as a mixer. I decided not to make use of the oscillator in this case. The NE612 operates from an 8V supply provided by IC2, a 78L08.

As mentioned previously, the output contains many signals, audio and RF. The RF signals are unwanted and are removed by RFC2 and C17. The higher audio frequencies are progressively attenuated by R9 and C18, which form a low-pass filter. This gives 3dB attenuation at 1.3kHz and 6dB per octave thereafter. TR4 is the audio preamplifier. A BC109 was chosen rather than a BC108 because it gives more gain, which is required here.

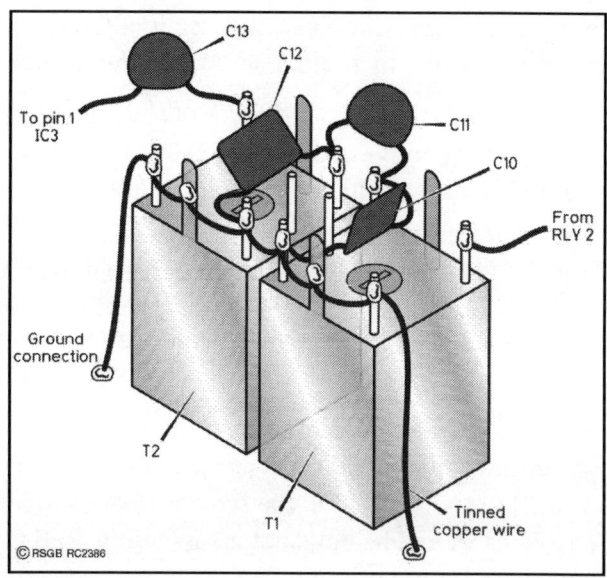

Fig 5. T1 and T2 are mounted upside down on the earth plane and wired as shown.

The audio output stage is an LM380N. As shown in **Fig 6**, it is mounted upside down in the 'dead bug' style.

Unlike the previous stages, it is supplied with 13V on transmit as well as on receive, as it is needed to amplify the sidetone signal to loudspeaker level.

TRANSMITTER

Referring to **Fig 7**, when the key contacts are closed, RLY2 switches the aerial to the PA stage and 13V from the receiver to the transmit circuits. The RIT control on the front panel becomes no longer operative, and the VFO offset voltage now comes from RV3.

The VFO signal is fed to the input of the driver stage, a BFY51. The collector is coupled to the PA by L2, a toroidal transformer. It is more usual to tune this with a trimmer but (on 80m at least) the tuning is very flat, so the number of turns is optimised for a standard-value capacitor, in this case 150pF (C29). There is adequate drive over the whole of the tuning range.

Low-frequency parasitic oscillations, normally below 100kHz, sometimes occur in the driver stages of QRP transmitters, and can go unnoticed by the operator. I remember coming across a rough CW signal on about 3630kHz. I identified the station concerned, tuned down the band and also found him on 3560kHz. The two signals were 70kHz apart and, as expected, I found another rough CW signal at 3490kHz. The station concerned was only running 3W and was nearly 200 miles away.

This problem is normally caused by poor decoupling and is easily cured. I have had no such problems with this design. The output from TR5 is taken via a link winding on L2 to the base of TR6, the PA. This is a BFY51 and produces in excess of 1W output. The DC feed choke, RFC3, is not critical.

Some variation in wire thickness and number of turns shouldn't cause

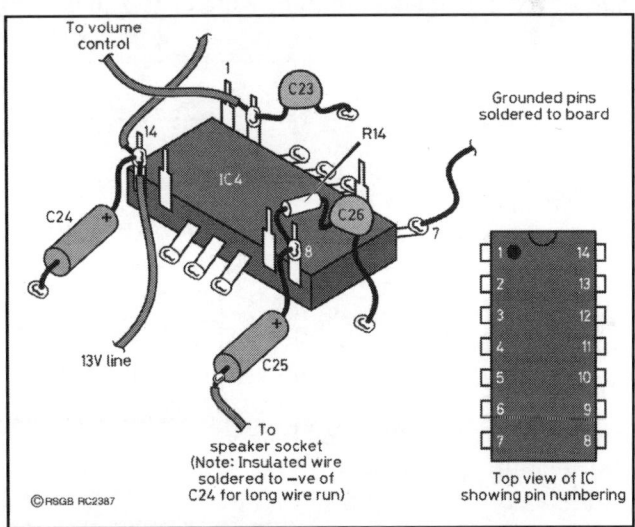

Fig 6. IC4, the audio output chip, is mounted 'dead bug' style on the ground plane and wired as shown.

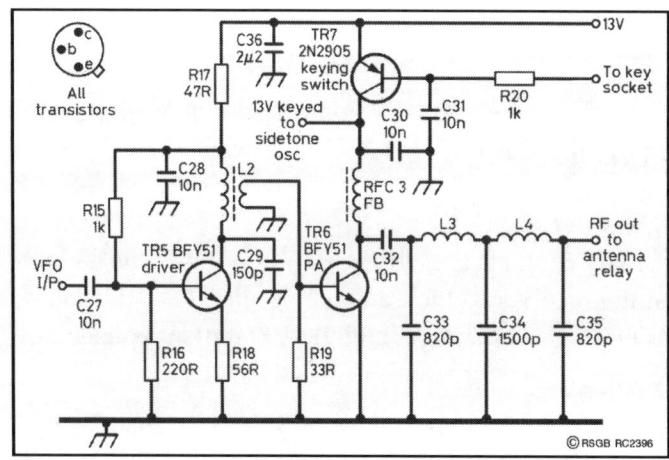

Fig 7. The transmitter produces over 1W output and, thanks to adequate decoupling, is quite stable.

any problems, so just try what you have available. Take care, though, not to damage the enamel insulation when passing the wire through the bead, as this is easy to do.

The output is coupled by C32 to the low-pass filter, comprising L3, L4, C33, C34 and C35. It then passes via the changeover relay RLY2 to the aerial socket.

TR7 is the keying switch. Its use is preferable to directly-keying an RF stage such as the PA or driver, which can be unpredictable because RF has a tendency to find its way into keying lines. Also, there is the added advantage that the key or keyer only has to cope with the base current of TR7, a few milliamps.

The disadvantage of this kind of circuit is that it appears to be very good at injecting RF into supply lines and causing a shift in VFO frequency on 'key down' – something which is all too common in simple QRP equipment. The solution, again, is adequate decoupling, which is taken care of by C30 and C31.

RIT

The requirement here is that the VFO frequency can be varied over a small range without moving the main tuning control. As you can see from **Fig 8**, RV2, the RIT control, is active on receive only. On switching to transmit, the RIT control has no effect and the frequency reverts to that set by the main tuning.

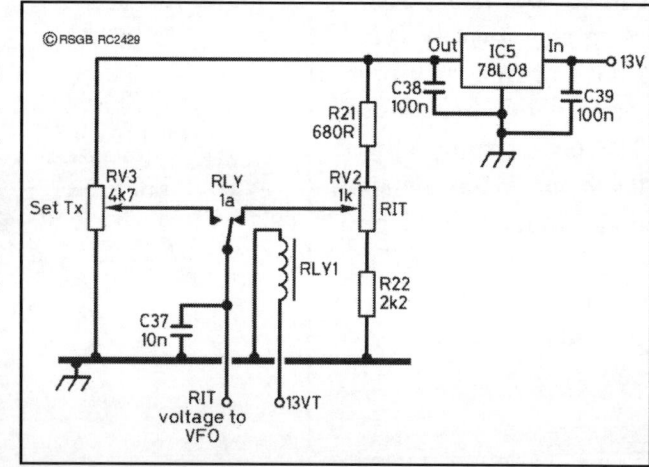

Fig 8. How the RIT voltage to the VFO is switched between transmit and receive.

The actual shift in frequency is accomplished by a BB409 varicap (variable capacitance) diode coupled to the VFO tuned circuit. When this is reverse-biased (ie positive to the cathode) it exhibits a capacitance which is dependent on the voltage so, by varying the voltage, it will tune the VFO over a small range. A fixed, stable voltage is required on transmit. This is provided by RV3. On receive, a variable voltage is provided by RV2, the front-panel-mounted RIT control. The supply to RV2 and RV3 is stabilised by IC5. Relay RLY1 selects the voltage from RV2 or RV3, depending on whether the transceiver is in the receive or transmit mode.

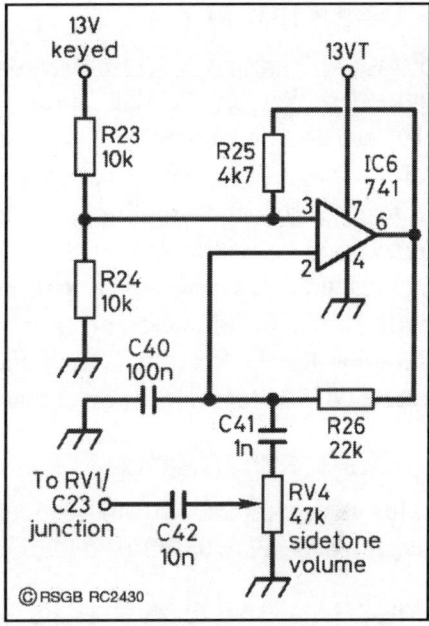

Fig 9. The sidetone oscillator and its associated level control.

SIDETONE

The sidetone oscillator (**Fig 9**) uses a 741 IC. It draws its supply from the 13V line, but is only activated on 'key down' by 13V from the keying switch, TR7. The output is taken to the volume control, RV1, via the sidetone level control, RV4, which is used to reduce the level of signal reaching the volume control. This is for operator comfort.

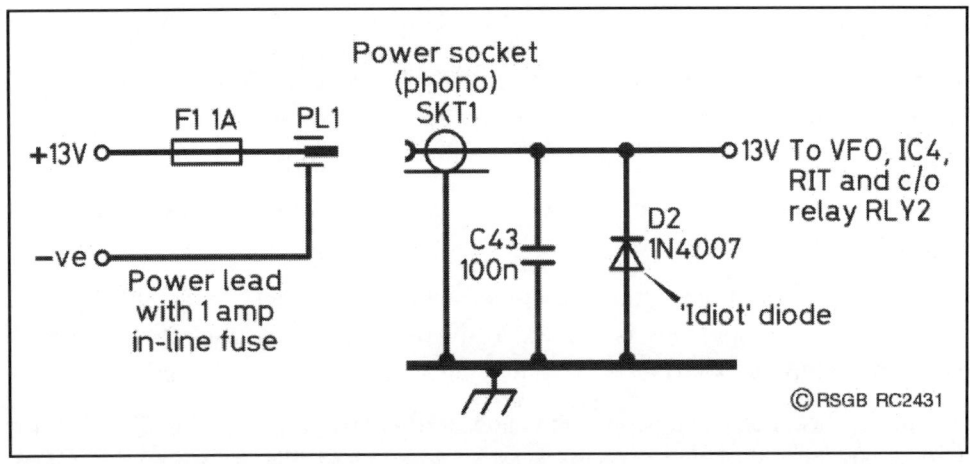

Fig 10. The connection to the supply contains protection against power of the wrong polarity being applied accidentally.

POWER SUPPLY

Power is supplied to the transceiver via a phono socket on the rear apron (see **Fig 10**). This is decoupled by C43. There is a diode, normally reverse-biased, connected directly across the supply input. This is for protection so, if the supply is accidentally connected the wrong way round, the diode will be forward-biased and will conduct, blowing the 1A fuse, which should be installed in a holder in the power lead.

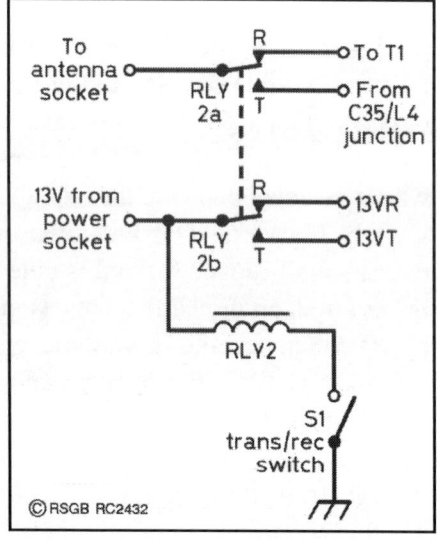

Fig 11. Transmit / receive switching.

AERIAL SWITCHING

This is accomplished using a double-pole changeover relay RLY2 (see **Fig 11**).

MECHANICAL CONSTRUCTION

As stated previously, this project was designed for easy construction. Mechanical work has been minimised; nonetheless some is required. The copper-clad board should be double-sided glass fibre but SRBP could be used. This would be easier to cut, but the end result is not as rigid. If you have, or if you know someone who has, access to a workshop guillotine, this part of the project could be very easy.

If not, you will have to cut the board by hand. Possibly the best way to do this is clamp it in a vice between two pieces of angle iron which are then used as a cutting guide.

Using a hacksaw, cut the pieces very slightly oversize, then file down to the required dimensions. Refer to **Fig 12** and **Fig 13** for the dimensions.

When you have cut the base, front panel, rear apron, VFO enclosure and the two triangular supports, the holes can be drilled. First, mark their locations accurately and use a centre punch.

Start by drilling all holes 3mm in diameter. The larger holes can then be drilled to 6mm. At 10mm, more care is needed, as the drill can easily bite into the PCB. If you can, clamp the board firmly and drill through as slowly as possible.

The aerial socket can be any type you choose. If you want to fit an SO-239, a 16mm hole will be needed. This can be made by marking out, drilling to 10mm and enlarging with a file. Alternatively, drill a series of small holes (say, 2mm) inside the circumference of a 16mm circle, cutting out the centre and finishing off with a file.

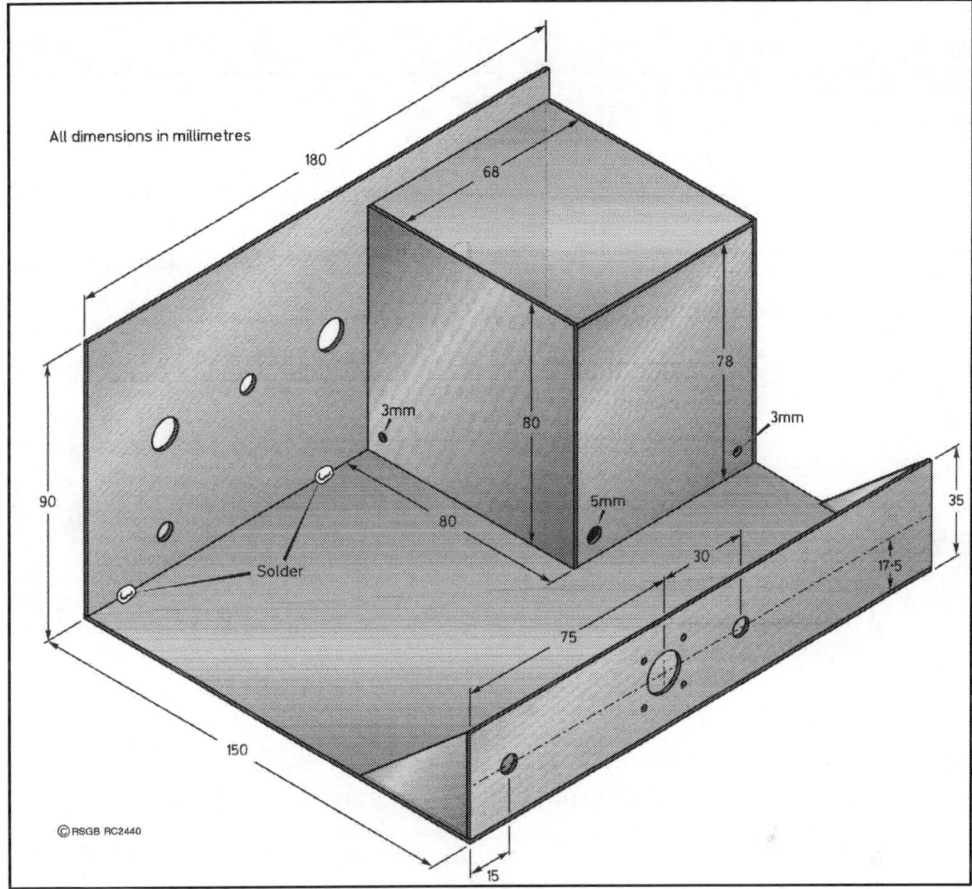

All dimensions in millimetres

180 · 68 · 78 · 3mm · 80 · 3mm · 90 · 5mm · 80 · Solder · 35 · 30 · 17·5 · 75 · 150 · 15

© RSGB RC2440

Fig 12. *The chassis, seen from the back.*

Finally, drill the two 3mm holes in the base for the screws which support the VFO board. In this instance, there is no need to measure – just place the matrix board that you will be using in position, and mark through the holes.

ASSEMBLY

With everything cut out and drilled, assembly can begin. Start by mounting the front panel on the base. Initially, use just one blob of solder in the centre. Then fix the side pieces of the VFO enclosure, again soldering lightly. After this, solder the rear apron in place. Finally, solder the two triangular supports in place. Do not mount the rear of the VFO enclosure at this stage. Inspect your work and if happy apply more solder to the joints and also solder in more places.

The end result will look more pleasing if the front panel is faced with card or thick paper, bearing labelling for the controls, along with a tuning scale for the

VFO, and zero and ± marks for the RIT control. This can be done at any stage and fixed with adhesive.

ELECTRICAL CONSTRUCTION

This is best done in a logical order, rather than haphazardly. Refer to **Fig 14**, and start by mounting the large components: controls, sockets, switches, relays (the latter can be fixed in place with a spot of Super Glue®). Draw pencil marks on the base, along the line

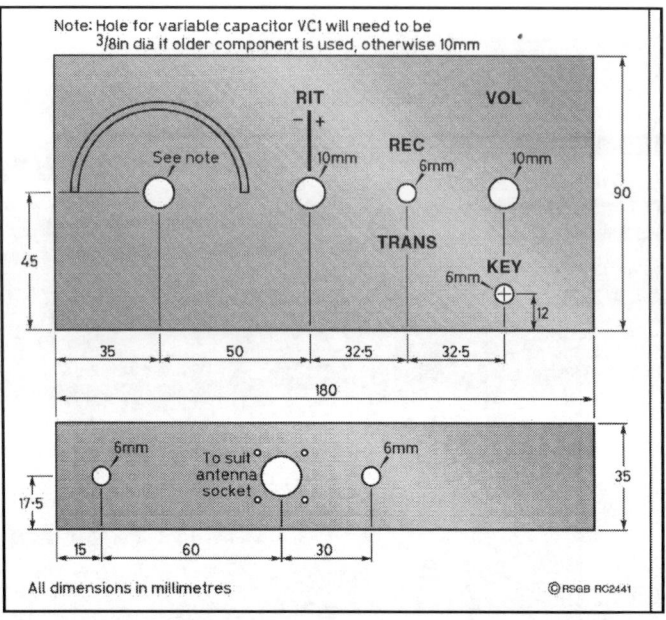

Fig 13. Front panel and rear apron layouts.

of the receiver (30mm from the side edge) and transmitter (30mm from the back edge). These are used as guides for the smaller components.

The first major job is the VFO. Fit all components and make all connections, taking great care to get the pin connections of the semiconductors correct. Check and double-check everything. Mount the VFO board, ensuring that the connections on the underside are clear of the base. Connect the variable capacitor using stiff wire (1.25 or 0.9mm). The lead to the RIT pin can be grounded for now. Connect a 13V supply and check that there is 5V at the output of IC1. All being well, the tuning range can now be roughly set. With VC1 at half-mesh, connect a short length of wire to the output (C9). Using a test receiver tuned to 3550kHz, adjust TC1 until a signal is heard. This is all that is needed at the moment. Don't be concerned if the VFO is not very stable at this stage.

Now for the receiver. Referring to **Fig 4**, start by assembling the band-pass filter (**Fig 5**) and fix it in place. Continue with IC3, IC2 and the smaller components. IC3 is mounted in the same manner as IC4 ('dead bug', ie legs upwards), but only pin 3 is grounded. Where grounding is necessary, leads are soldered directly to the base. Carry on until the receiver is complete. The volume control is connected with miniature screened audio cable, and the output from IC4 is connected to the loudspeaker socket. The receiver can now be tested using a temporary connection

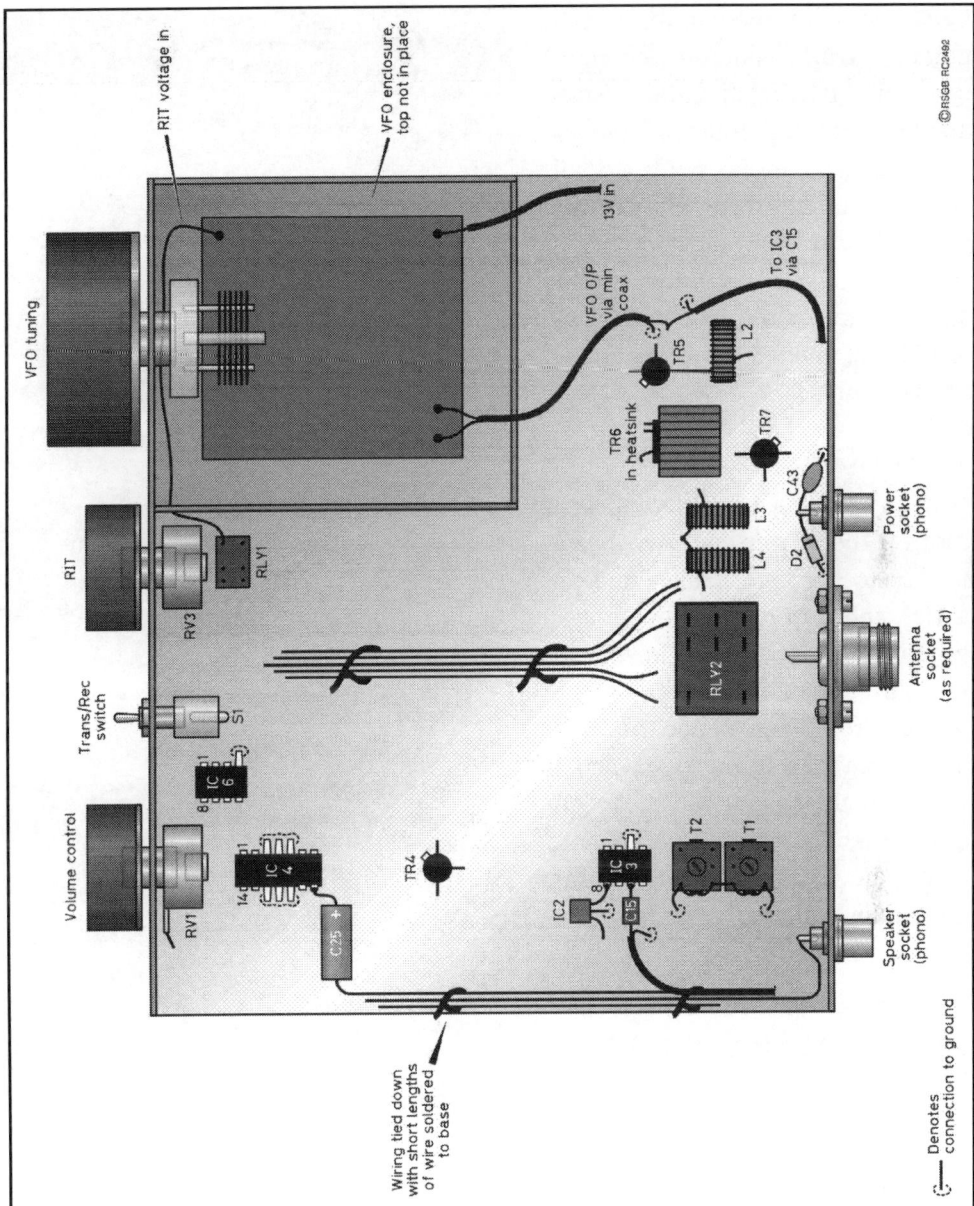

Fig 14. *Layout of the major components. Smaller components are soldered between the larger ones.*

from the VFO to IC3, with 13V to IC4, TR4 and IC2, and a length of wire as an aerial to the input of the band-pass filter. Sensitivity will be low, but it should be possible to hear something by tuning the VFO up and down its range.

Continue with the auxiliary circuits – the RIT, sidetone, change-over relay RLY2 [1] and the power connector. Permanent power connections can be made to all stages, a length of miniature coax connected to C15, and the wiring tied down neatly as shown. Now carry on and complete the transmitter and make a thorough check of everything.

TESTING

If at any stage a problem appears, turn off the power and investigate it. Connect a loudspeaker and an aerial, and turn on. Advancing the volume control should result in noise from the loudspeaker. Set the RIT control to zero. Tune the VFO and search for a steady signal at the centre of the tuning range. When one is found, using a suitable tool, adjust the cores of T1 and T2 for maximum volume. Now check the function of the RIT control.

Disconnect the aerial and connect a Morse key and a dummy load [2]. Set RV3 to the centre of its travel. Switch to transmit and press the key. The sidetone should be heard in the speaker. Holding the key down, tune the test receiver until the signal is found, somewhere near 3550kHz. If you have a power meter, it can be connected between the transceiver and dummy load. It should show an output of at least 1W.

Table 1. RIT, sidetone, switching and power input circuits components list

Resistors

R21	680R
R22	2k2
R23, 24	10k
R25	4k7
R26	22k
RV2	1k
RV3	4k7
RV4	47k

Capacitors

C37, 42	10n ceramic disc
C38–40, 43	100n ceramic disc
C41	1n ceramic disc

Semiconductors

IC5	78L08
IC6	741
D2	1N4007

Miscellaneous items

RLY1	12V single-pole change over relay
RLY2	12V double-pole change over relay
F1	1A fuse with in-line holder
S1	SPST switch
PL1	Phono plug
SKT1	Phono socket

Table 2. Transmitter components list

Resistors

R15, 20	1k
R16	220R
R17	47R
R18	56R
R19	33R
All resistors 0.6W metal film	

Capacitors

C27, 28, 30–32	10n
C29	150p
C33, 35	820p
C34	1500p
C36	2µ2

Inductors

L2	42 turns of 0.375mm enamelled wire on a T50-2 core, plus a 5-turn link of PVC-covered wire over the main winding
L3, L4	22 turns of 0.56mm wire on T50-2 core
RFC3	20 turns of 0.19mm wire on ferrite bead

Semiconductors

TR5, 6	BFY51

Table 3. Receiver components list	
Resistors	
R9	5k6
R10	120k
R11	33k
R12	3k3
R13	1k
R14	1R2
RV1	22k log pot
All resistors 0.6W metal film unless specified otherwise	
Capacitors	
C10, 12	47p ceramic plate
C11	4p7 ceramic plate
C13	100p ceramic plate
C14, 17	10n ceramic disc
C15	22p polystyrene
C16, 22	1μ, 16V electrolytic
C18	22n ceramic disc
C19	10μ, 16V electrolytic
C20, 21	10n ceramic disc
C23	1n ceramic disc
C24	47μ, 16V electrolytic
C25	100μ, 16V electrolytic
C26	100n ceramic disc
Inductors	
RFC2	1mH
T1, 2	Toko KANK3333R
Semiconductors	
TR4	BC109
IC2	78L08
IC3	NE612
IC4	LM380N

Table 4. VFO components list	
Resistors	
R1, 2	100k
R3	15R
R4	820R
R5	1k5
R6	33R
R7	12k
R8	270R
All resistors 0.6W metal film	
Capacitors	
C1	10p ceramic plate
C2	150p close tolerance polystyrene
C3, 4	750p close tolerance polystyrene
C5	22p polystyrene
C6, 7	100n ceramic disc
C8, 9	10n ceramic disc
VC1	10 or 12p (see text)
TC1	65p (Maplin WL72P)
Inductors	
L1	50 turns of 32SWG on T50-2 toroidal core
RFC1	1mH
Semiconductors	
TR1	2N3819
TR2, 3	BC108
IC1	78L05
D1	BB409

SETTING UP

There are only three adjustments to be made – the VFO, RIT and bandpass filter – and they have already been roughly set.

VFO

Close the vanes of VC1 fully. The tuning knob should be set to exactly nine o'clock. Now set the test receiver to 3490kHz (if VC1 is a 12pF component) or 3500kHz if it is 10pF.

Adjust TC1 until the signal is heard. Set the tuning control to the three o'clock position and check the frequency of the VFO by finding the signal with the test receiver. If using 10pF for VC1, the frequency should be about 3600kHz; if using 12pF it should be about 3610kHz. The VFO enclosure can now be completed, but don't use too much solder on the lid – you might want to remove it at some future time. The

VFO must be calibrated, but not yet. Since you have just heated everything up with a soldering iron, now is not a good time – allow several hours to elapse first.

RIT

Set the VFO to the centre of its range, the RIT control to zero (centre), and tune the VFO to give a beat note on the test receiver. Switch to transmit and adjust RV3 for exactly the same note. Switch between transmit and receive and carefully adjust RV3 so that the note doesn't change.

Bandpass filter

With the aerial connected, find a steady signal in the centre of the tuning range and carefully adjust T1 and T2 for maximum signal strength. This time take care and make sure it is right.

Calibration

It just remains to calibrate the VFO scale and the transceiver will be ready for use.

Before using it in earnest, though, it would be a good idea to get a local amateur to listen to your signal to make sure that all is well.

AND FINALLY...

The transceiver is delightfully easy to use. To net onto a station, start with the RIT at its centre position and adjust the main tuning for zero beat. The RIT can now be set to give the desired note.

With a reasonable aerial (eg a G5RV) you shouldn't have any shortage of QSOs. You are unlikely to achieve WAC or DXCC (please prove me wrong) – on the other hand, you will almost certainly have no problems with TVI or BCI. Have fun!

NOTES

[1] As a particular relay is not specified for RLY2, you will have to work out the connections for yourself. If no data are available, this can be done visually and confirmed with a test meter on a resistance range.

[2] This can be three 150Ω 0.5W resistors in parallel.

Earth-Continuity Tester

When using mains-powered electrical equipment, a good-quality protective earth system is very important for safety. Good earth connections are additionally important for radio operation, both for protection against lightning strikes and also for the greater effectiveness of antennas that use earth as one half of a dipole.

In situations where the earth path is a functional earth as opposed to a protective earth, a simple low-voltage, low-current continuity tester or resistance meter is usually sufficient for checking earthing resistance, but for a proper test of a protective earth a high-current tester is needed. This is because a deteriorating earth connection in the form of a stranded wire where many of the strands are broken will still show a low resistance to a low-current tester but, in a fault situation when the earth path needs to pass a high current to ground and thus trigger a protective device, the high current causes the remaining strands to 'burn out', ie go open-circuit, before the protective device has time to operate; the protection is then nonexistent.

SAFETY STANDARDS

Recognising this situation, the British and European safety standards for electrical safety, for example BS EN 60335-1 for household equipment, demand that the resistance of the protective earth path between an exposed metal part and the protective earth pin is less than 0.1Ω. The equipment needed for checking to this standard is specialised and expensive, but this simple project provides a low-cost alternative and will check resistance at 2–3A if good-quality batteries are used.

To simplify use, the circuit gives a pass/fail indication instead of a resistance value.

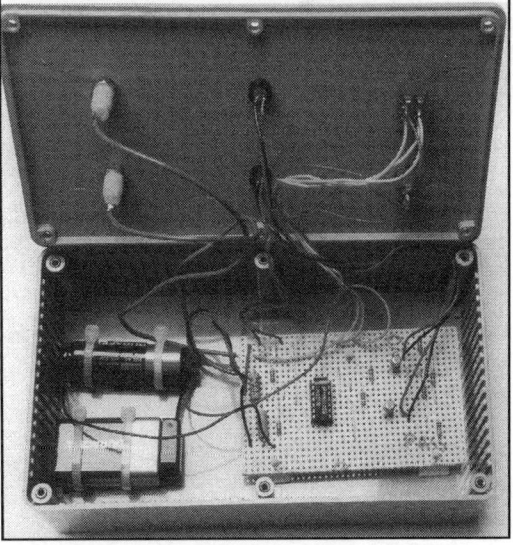

Inside a completed tester. Note that in a 'cased' project, the LEDs are removed from the stripboard and brought out to the front panel

WHEATSTONE BRIDGE

The circuit can be considered in

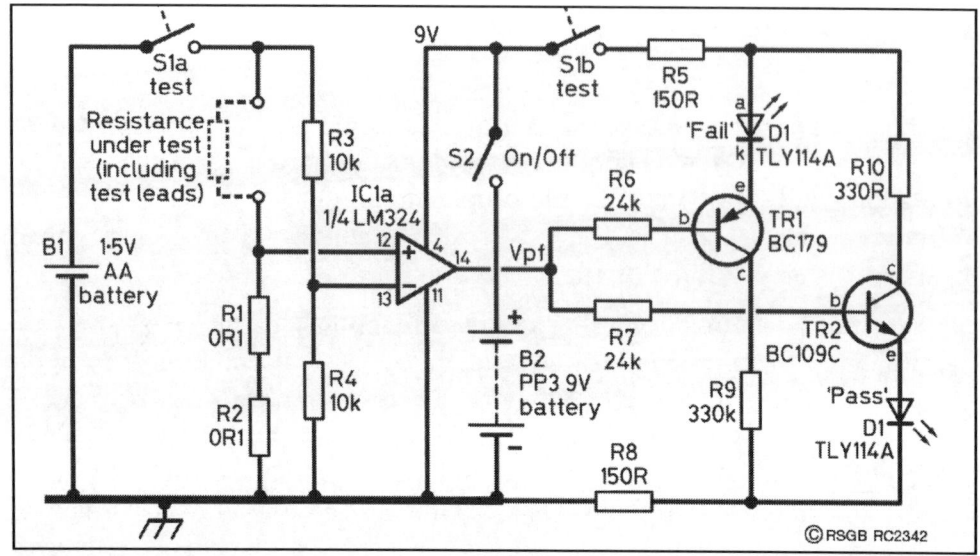

Fig 1. *Circuit diagram of the earth continuity tester*

three parts; *test, detector* and *output indicator. See* the circuit diagram in **Fig 1**. The test part of the circuit is based on a Wheatstone bridge, where the earth resistance path forms one of the resistance 'arms'. *See* **Fig 2** for the principle behind a Wheatstone bridge. As a consequence of the values of resistance chosen (the test leads are assumed to have a resistance of 0.1Ω), if the earth resistance is less than 0.1Ω, the voltage between the midpoints of the two halves of the Wheatstone bridge will be positive, and if it is less than 0.1Ω, it will be negative. This is fed to the detector part of the circuit.

The *detector* is an op-amp wired as a comparator. Connected in this way, it has such a high gain that its output is roughly equal to either the positive or negative supply rail voltage, depending on whether the PD between its non-inverting and inverting inputs is positive or negative. It doesn't matter whether the PD is large or small – the output will always be at either extreme. This means there will always be a definite pass or fail indication from the detector, no matter how large or small the output from the Wheatstone bridge. This is important, as it means correct operation of the circuit doesn't depend on the voltage of the high-current battery, particularly as it is a chemical type whose output voltage can fall dramatically when a high current is being drawn. The pass/fail voltage Vpf from the detector then passes to the output indicator circuit

The *output indicator* circuit consists of two LEDs, driven by transistors to provide sufficient current, which indicate either a pass or a fail for an earth path resistance of less than or more than 0.1Ω. TR2 is an npn transistor which switches on when its input is high, while TR1 is a pnp type which switches on when its input is low. A separate supply voltage is needed for the op-amp and LED circuit, since the test battery voltage will drop under a heavy load current. Because the output of the op-amp does not swing completely to the positive and negative supply rails, measures need to be taken to ensure that the LED driver transistors switch off correctly.

CONSTRUCTION

A suitable stripboard layout is shown in **Fig 3** on the previous page, and **Fig 4** shows how to identify and orientate several of

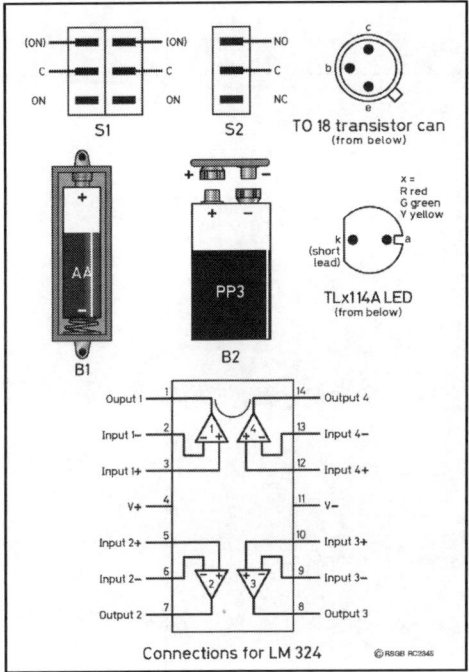

Fig 4. *Orientation (and pin-outs) of the batteries, IC1, LEDs and transistors*

the components. Use thick wire for the test leads!

HOW TO USE

Using flying leads with suitable connectors, eg crocodile clips, connect the circuit to each end of the earth path to be tested. This would usually be the mains plug earth pin and any metal part meant to be earthed. Then press the test button. Release the test switch as soon as a pass/fail indicator lights (certainly within 5 to 10 seconds, to lengthen battery life and prevent possible overheating of R1 and R2).

SAFETY NOTICE

The project described here may be used to test the resistance of appliance earth connections, but it is not intended to conform to any legal requirements for the testing of electrical safety

Table 1. Components list	
Resistors	
R1, 2	OR1, 2.5W
R3, 4	10k
R6, 7	24k
R5, 8	150R
R9, 10	330R
All resistors metal oxide 0.4W 1%, except R1 & R2	
Semiconductors	
IC1	LM324
D1, 2	TLY114A yellow, or TLR114A red and TLG114A green
TR1	BC179 (general-purpose pnp)
TR2	BC109C (general-purpose npn)
Additional items	
B1	1 × AA Duracell
B2	PP3
S1	Double pole, momentary on, or push-to-make
S2	SPST
Battery clips/holders	
Stripboard	
Plastic case*	
2 × 4mm plugs & sockets*	
2 × crocodile clips	
*Only required if you are building the project in a case.	

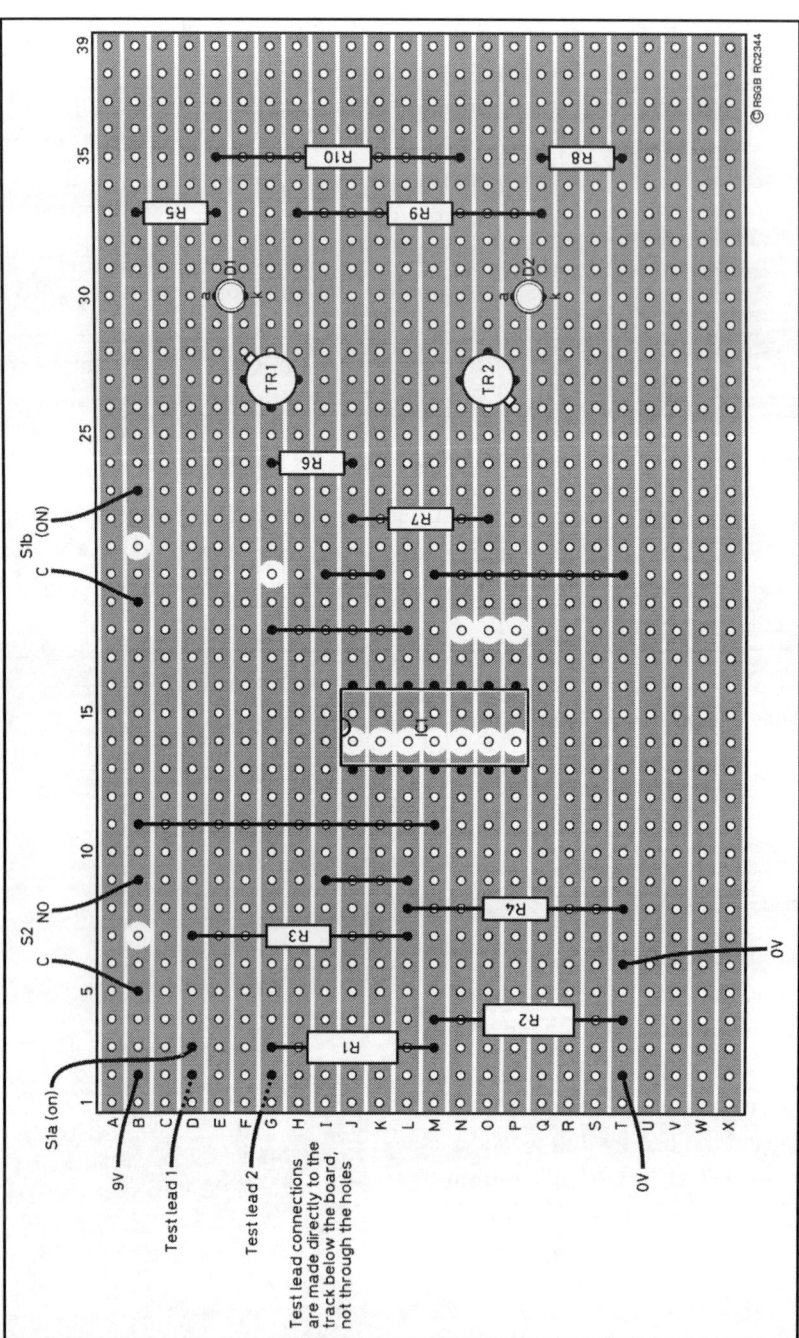

Fig 3. *Layout of the circuit on stripboard*

The RSGB and the author accept no responsibility for any accident or injury caused by its use. Never use on mains equipment plugged into the mains – the connection to the mains plug earth pin mentioned in the previous paragraph implies that the plug is free – *Ed*

Dry Battery Tester

The output voltage of a dry battery with no load, as measured by a digital voltmeter for example, will give an indication of the state of the battery. However, that is only part of the story – even a spent battery can show quite a high voltage when out of circuit but, as soon as an attempt is made to draw current, the voltage will fall dramatically.

You can check the current capability of a dry battery by connecting it directly across an ammeter but, as this is effectively short-circuiting it, it is not to be recommended. It not only wastes the energy of a good battery but can also cause overheating and other permanent damage if the short-circuit lasts for more than a second or two. A better method is to connect the battery to a load similar to the one it would experience in normal use and measure the current the battery is capable of providing. The device described here uses this method, and will test the current capability of the two most commonly-used dry cells, the 1.5V and 9V types, and provide a visual indication of the state of the battery via a series of LEDs

HOW IT WORKS

For the test, the battery is connected to a load resistor capable of withstanding, without 'burning out', the current the battery can provide. The voltage dropped across the resistor, which is proportional to the current (by Ohm's law), is compared with a series of reference voltages chosen to represent the changeover points between the current ranges of interest. Where the generated voltage exceeds the particular reference voltage, one or more of a series of LEDs light to give an indication of how much current

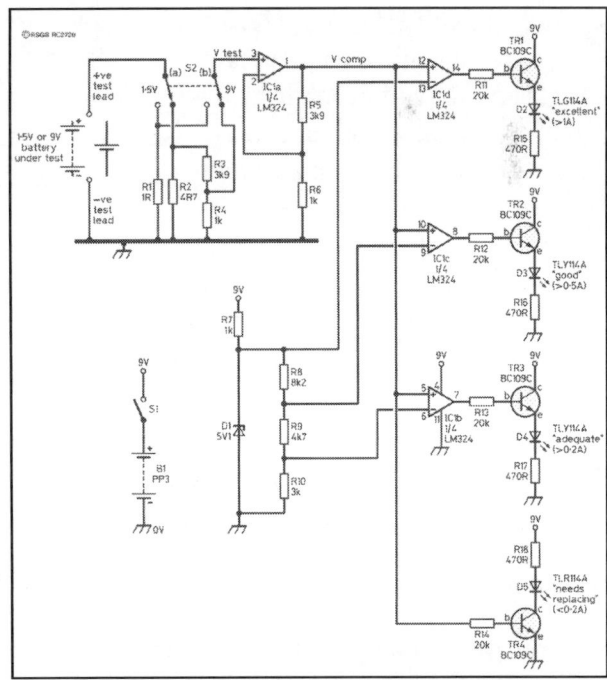

Fig 1. The unit works by checking the voltage that a battery is able to deliver when a substantial load is placed on it.

the battery is supplying and hence whether it is in an 'excellent', 'good', 'adequate' or 'needs replacing' state.

THE LOAD

AA- and PP3-type Duracell batteries will provide up to 2A for intermittent use. A suitable test load that draws sufficient current to test the battery's state without wasting too much of its power will be one that allows around 1A to flow.

THE CIRCUIT

This is shown in **Fig 1**. R1 and R2 are the load resistors for 1.5V and 9V batteries respectively, and switch S2(a) selects whichever test is required. The same value of current passing through a 1Ω and a 4.7Ω resistor will generate a different potential difference (PD) across each resistor of course, and so for the indicator part of the circuit to be common to both tests,

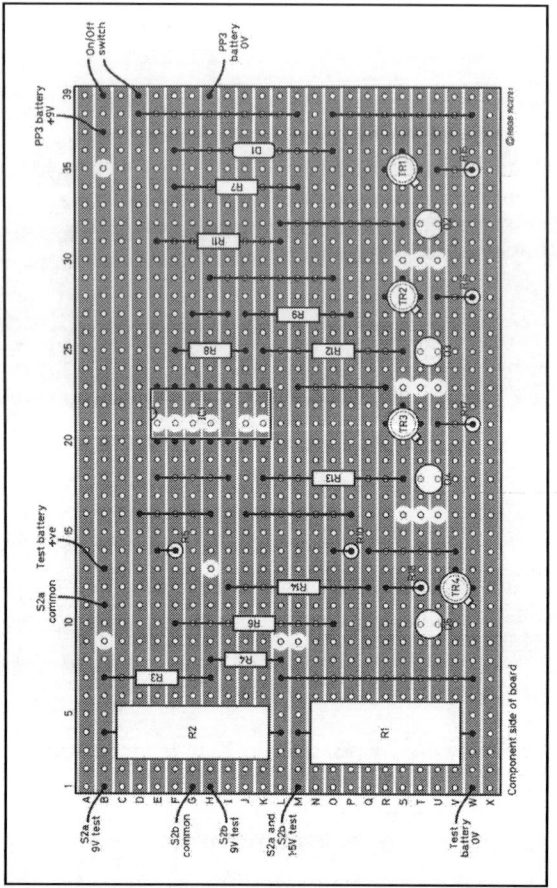

Fig 2. Stripboard layout of the dry battery tester.

the PD across the 4.7Ω resistor is split by a potential divider consisting of R3 and R4. Using this method, the value of Vtest is nominally 1V/amp of current supplied by the battery under test for both types of battery. S2(b) selects between the non-divided and divided source voltage as appropriate.

Before passing to the comparator stage, Vtest is amplified by IC1a, connected as a non-inverting amplifier with a nominal gain of five. This means that the voltage applied to the comparators (Vcomp) will range from 0V to around 6V, and give more predictable switching of the comparators by avoiding using voltages around or below the turn-on voltages of the IC's internal p-n junctions (about 0.7V).

The reference voltages against which Vcomp is compared are generated by a voltage divider network consisting of R8, R9 and R10 across a 5.1V Zener diode, D1. The use of a Zener diode ensures that the reference voltages remain stable,

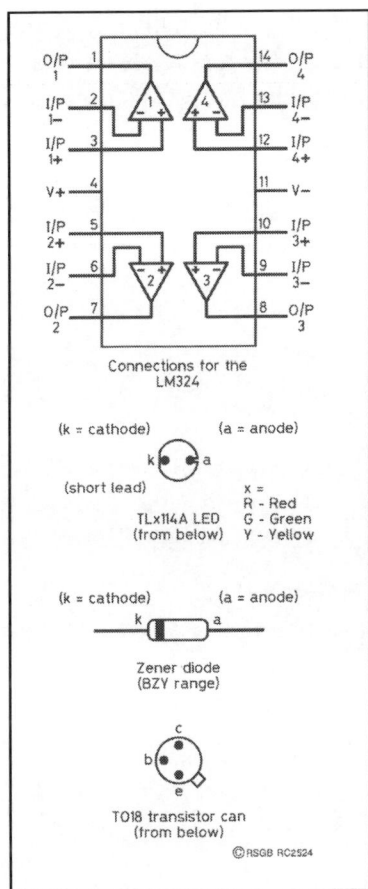

Connections for the
LM324

(k = cathode)   (a = anode)

k   a

(short lead)
                              x =
                              R - Red
TLx114A LED   G - Green
(from below)   Y - Yellow

(k = cathode)   (a = anode)

k   a

Zener diode
(BZY range)

c

b

e

TO18 transistor can
(from below)

© RSGB RC2524

Fig 3. Orientations of the semiconductors.

even if the power supply battery voltage changes with use. R7 limits the current through D1 to its working value. Reference voltages of 1V, 2.5V and 5.1V are used to give changeover points at nominally 0.2A, 0.5A and 1A.

Indication of current values of less than 0.2A (but greater than approximately 0.14A) are given by the LED driven by TR4. This part of the circuit takes further advantage of the turn-on voltage of a p-n junction, in that TR4 will not switch on until Vcomp is above around 0.7V. Hence there will be no false indication of a current of less than 0.2A when no battery is being tested, even if Vcomp is not exactly zero due to any small offset or bias currents associated with the op-amps. IC1b, IC1c and IC1d are op-amps used to act as comparators by not employing any feedback. Due to the high gain of op-amps used in this way, their output will be around 0V when the non-inverting (+) input is less than the reference voltage on the inverting (−) input. In this state, the transistor connected to its output (one of TR1 to TR3) will be 'off' and the associated LED (one of D2 to D4) will not light. As soon as Vcomp rises above the relevant reference voltage however, the op-amp output will immediately go to around 9V, switching 'on' the appropriate transistor and lighting the associated LED. R11 to R18 limit the current to the working values of the transistors and LEDs.

CONSTRUCTION

A suggested stripboard layout is shown in **Fig 2**. Construction is straightforward, the only particular point to mention being the correct orientation of the IC, transistors, Zener and LEDs. The markings that indicate their correct orientation are reproduced in **Fig 3**.

The components are all general-purpose low-power types, apart from the load resistors which ideally need to dissipate 5W (although 3W types should suffice as the current to be carried only passes for short periods). It is useful to employ different colours for the positive and negative test leads as the correct test-battery polarity must be observed for the LEDs to light correctly. Alternatively, solder battery cell holders and clips permanently to the ends of the test leads (if this is done only one test battery should be connected to any holder or clip at a time, and it should be removed immediately after testing).

BATTERY TESTING

With S2, select the 1.5V or 9V battery test as appropriate, switch on, connect the battery to the test leads and note which LEDs light

PARTS LIST

Resistors - All resistors metal film, 0.6W, 1%, unless specified otherwise

R1	1Ω, 5W, see text
R2	4.7Ω, see text
R3, 5	3.9k
R4, 6, 7	1k
R8	8.2k
R9	4.7k
R10	3k
R11 – 14	20k
R15 – 18	470Ω

Semiconductors

IC1	LM324
TR1 – 4	BC109C

D1	BZY5V1 0.5W Zener
D2	TLG114A green LED
D3, 4	TLY114A yellow LED
D5	TLR114A red LED

Additional items

S1	SPST
S2	DPDT

Stripboard
PP3 battery and clip
Battery holders for test batteries, if required, see text
Case to suit

A 'Loop' Alarm
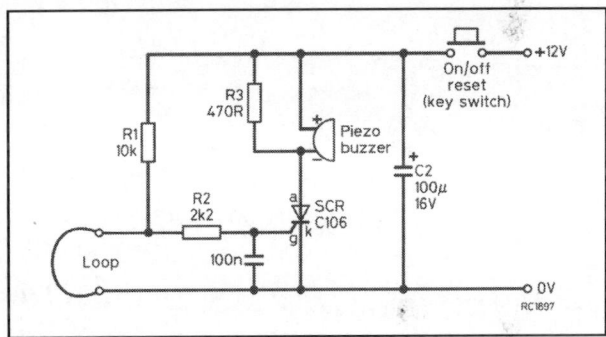

4.7

Putting on a station at a local fair or similar function is a good way of gaining publicity for our wonderful hobby but, when you are on a stand, it is not always possible to keep an eye on the equipment. This alarm is very useful for those situations. It is similar to those used in shops, where a loop of wire is wrapped around the goods. If the wire is broken to remove an item, the alarm sounds.

HOW IT WORKS

The alarm uses a thyristor, which is also known as a silicon controlled rectifier (SCR). [To avoid any misconception by reading the name wrongly, this is a rectifier that can be controlled, and it is made out of silicon – *Ed.*] An SCR is like any other diode in that it will only allow current to flow in one direction, ie from the anode, 'a', to the cathode, 'k'. However, unlike any other diode, this current will only start to flow when a small positive voltage is applied to the gate, 'g'. Having started to conduct, the SCR continues to

Fig 1. The loop security alarm uses the wire loop to prevent the SCR from being turned on.

do so, even if the gate voltage is removed. In the circuit, shown in **Fig 1**, the SCR does not conduct under normal conditions because the wire loop maintains zero volts on the gate. If the loop is broken, the gate is pulled positive by R1 and R2, which makes the SCR conduct, sounding the alarm. Even if the loop is re-joined, the SCR continues to conduct and the alarm continues to sound until the power is turned off.

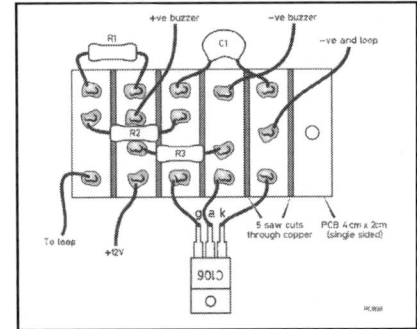

CONSTRUCTION

The circuit is built on a small piece of single-sided PCB. Saw cuts are made through the copper, as shown in **Fig 2**, to form pads.

Fig 2. Construction is simple on a small piece of PCB, but be careful when cutting through the copper not to cut right through the board!

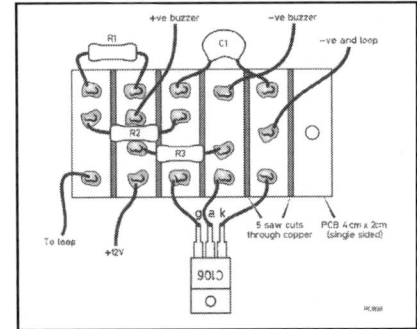

The components are soldered to these pads. The on/off switch should ideally be a key-switch, to prevent the alarm being switched off by an unauthorised person.

The ends of the loop can be connected via practically any type of connector to the case containing the alarm; indeed, the loop could comprise of any number of short lengths of wire, joined with in-line plugs and sockets. The latter would allow a single item of equipment to be removed without disturbing the whole loop.

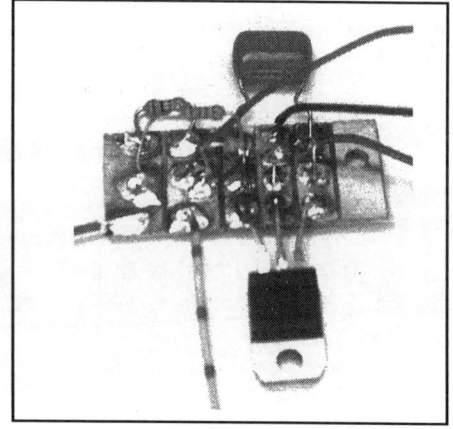

The completed project. In this photo the key-switch cannot be seen, as it is on the far side of the box.

PARTS LIST

Resistors - All resistors 1/4W

R1	10k
R2	2.2k
R3	470

Capacitors

C1	100n
C2	100µ electrolytic

Semiconductor

TH1	C106 (Maplin)

Additional items

Key-switch
Piezo buzzer (Maplin)
Plastic case
PCB material
Wire for loop
Plug(s) and socket(s) for loop

A Portable Power Supply

When time permits, I enjoy participating in the Backpacker series of RSGB contests [1], operating in the 3W category. The power supply I use for these outings is described here. It is ideal for the purpose, and can also be used as a simple uninterruptible power supply (UPS) for the shack.

DESIGN CRITERIA

- A maximum of 4.5kg (10lb) in weight.
- Able to be carried in a small rucksack.
- Able to be plugged into the nearest mains outlet to be recharged.
- When at home to run in 'float-charge mode', to operate low-current equipment.
- To be of reasonable cost.
- To use readily available components.

At the 1998 Rainham Rally I found a couple of sealed lead-acid cells rated at a nominal 12V and 7Ah capacity. They weighed in at a little over 2.2kg (5lb) – just what I needed!

Being of the sealed variety, care has to be exercised in not overcharging the cell, ie to prevent gassing, so 13.8V would be the maximum permitted voltage at the terminals. Thus a stabilised supply was essential.

THE CIRCUIT

The float charger is hardly original – indeed it was adapted from that excellent series of articles by John Case [2]. One major consideration was the need to 'over-engineer' it, as I would leave it plugged in and switched on almost continuously.

Fig 1 shows the circuit diagram. Many of the components were salvaged from redundant equipment or the junk box; however, the critical components – transformer, pass transistor, reservoir capacitor and regulator chip – were all purchased new.

There are many options available for layout; mine were dictated by the size and shape of the heatsink. In my case, the box that holds it all is made from 12mm plywood, with the major components mounted on an L-shaped aluminium plate which forms the front and part of one side.

A voltage-dependent resistor (VDR) is fitted across the primary of the transformer, and an over-current control is provided to limit the current to 1.5A.

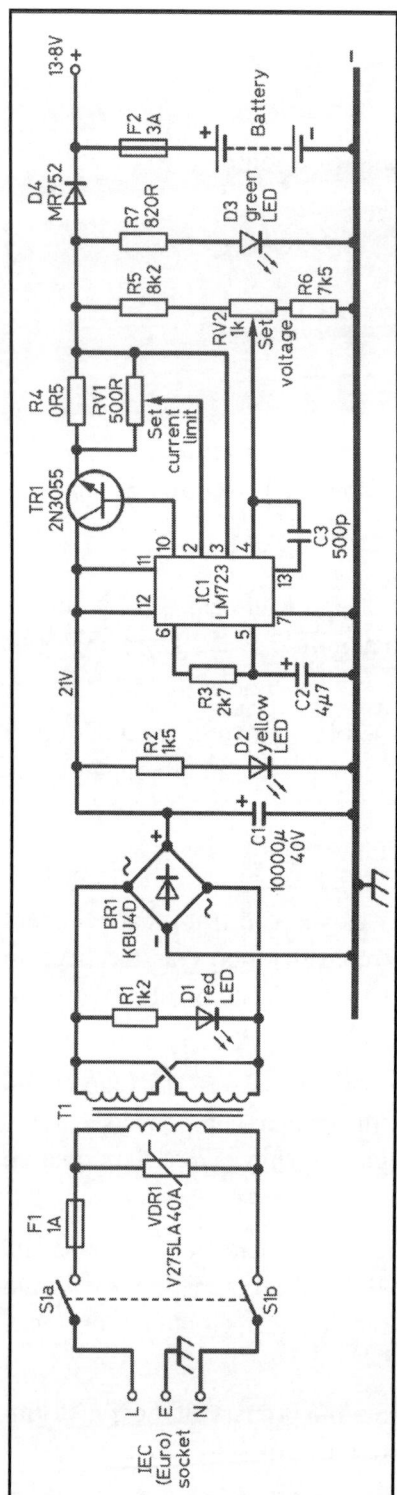

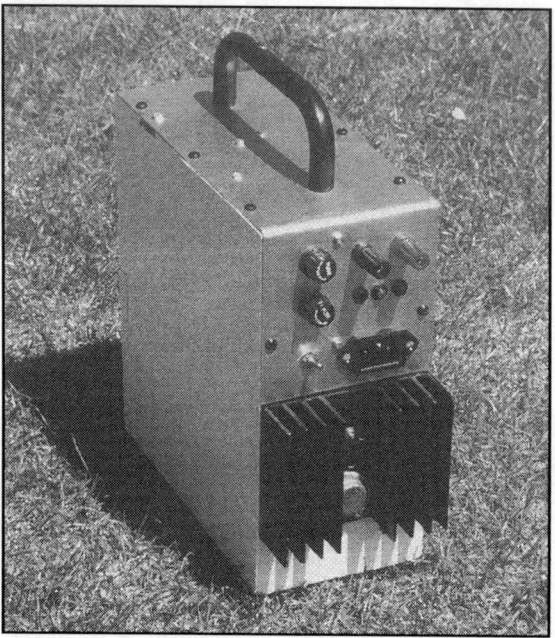

Fig 1. The portable power-pack works by float-charging a sealed lead-acid battery, to provide an uninterruptible supply.

The completed power supply.

A heavy-duty diode D4 is incorporated in the feed to the battery, and LEDs are fitted at strategic points as a confidence feature and for ease of fault-finding. The voltage controller, an LM723, is fitted, along with its components, on a small piece of matrix board, the remaining components being wired 'point-to-point' using substantial cable for the heavy-current paths.

Setting-up is quite straightforward – connect everything together to a fully-charged battery, connect a voltmeter to the battery's terminals, turn on and adjust the output to 13.8V as measured at the battery terminals. Set the current limit to 1.5A, which in practice appears rarely achieved.

RESULTS

Does it meet the criteria I set out earlier? That I will leave you to judge. The figures are:

- Weight – 11lb.
- Size – 25 × 25 × 10cm, including heatsink and handle.
- Cost – around £20.
- Capacity – enough for a full Backpacker session

Even when left on continuously, the temperature of the portable power pack hardly rises above room temperature.

PARTS LIST

Resistors - All resistors ½W metal film, 5% tolerance, unless otherwise stated

R1	1.2k
R2	1.5k
R3	2.7k
R4	0.5, 2W
R5	8.2k
R6	7.5k
R7	820
RV1	500 linear preset potentiometer
RV2	1k linear preset potentiometer
VDR1	V275LA40A

Capacitors

C1	10,000µ electrolytic, Maplin LE03
C2	4.7µ electrolytic
C3	500p ceramic

Semiconductors

BR1	KBU4D (or similar)
D1	Red LED
D2	Yellow LED
D3	Green LED

D4	MR752 (or similar)
IC1	LM723
TR1	2N3055

Additional items

F1	1A plus holder
F2	3A plus holder
S1	Double pole, single throw (DPST) toggle
T1	Mains transformer with two 15V @ 0.75A secondaries (Maplin DH27)

7Ah sealed lead-acid battery

IEC socket

Matrix board

Screw terminals, insulated

Case to suit

REFERENCES

[1] 'Backpacking – summertime delights', G6TTL, RadCom May 1997.

[2] 'Power supplies on a shoestring', GW4HWR, RadCom July – August 1986.

Homebrew Microphone

The microphone is one of the most important components of a telephony transmitter. As the microphone is the very first stage in a phone transmitter, any distortion or noise from the microphone or its amplifier is sure to degrade the quality of the transmitted signal. The amateur radio shack can be a harsh environment for a microphone. RF signals at high field strength and strong magnetic fields from power supply transformers are easily picked up by the mic. Acoustic noise from cooling fans is another source of unwanted noise.

A microphone is a transducer that converts sound (vibrations in air) into a varying electrical voltage or current. This electrical signal can be amplified and used to modulate a speech transmitter. 'Dynamic Mike' is not necessarily the name of a particularly lively radio amateur; the word 'microphone' is often abbreviated to 'mike' or 'mic'. There are many different types of microphone in common use.

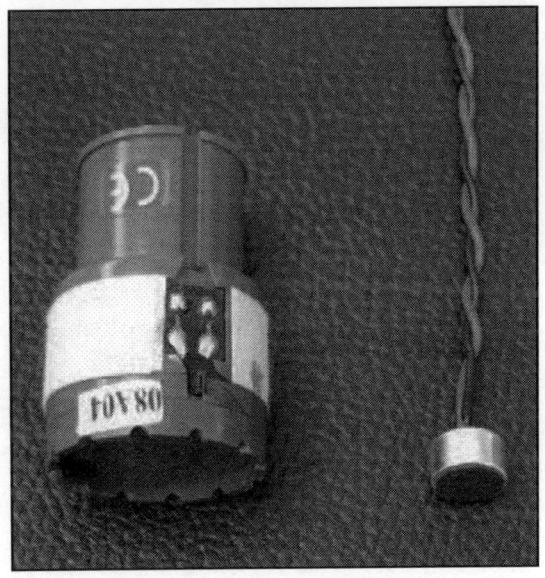

Photo 1: *Typical dynamic microphone (left) and electret condenser mic (right). The electret mic is 9.8mm diameter.*

- Some older military equipment used carbon microphones of the type that used to be found in telephones. The carbon microphone uses a loosely packed capsule of carbon granules that are in close contact with the diaphragm. Vibration of the diaphragm causes small changes in resistance of the carbon granules.

- Crystal microphones have a piezoelectric material in contact with the diaphragm. Stresses caused by the vibrating diaphragm generate a voltage across the piezoelectric material.

- Condenser (capacitor) microphones use one plate of a capacitor as the diaphragm. The second plate is fixed in position. A polarising bias voltage is applied across the two plates. Movement of the diaphragm plate relative to the fixed plate causes the capacitance to vary. This produces a changing output voltage.

• The dynamic microphone is an electromagnetic device that works in the same manner as a moving coil loudspeaker. The diaphragm is connected toacoilthatissuspendedinamagnetic field. Movement of the diaphragm will produce an electric current in the coil.

Carbon microphones are rarely used today because they tend to suffer from higher levels of distortion than other microphone types. The carbon mic is also very sensitive to humidity. Crystal microphones are not as popular as they used to be, but you will still hear many older examples on the amateur bands today. The two most popular types of microphone in use today are the dynamic mic and one particular type of condenser microphone known as the electret condenser mic. Both types are inexpensive, readily available and capable of producing high quality audio.

CONDENSER MICROPHONE

Fig 1 shows the configuration of a condenser microphone. The two capacitor plates are on the left of the schematic. The fixed plate is connected to ground; the diaphragm plate is connected to a polarising voltage supply. Movement of the diaphragm varies the capacitance by changing the spacing between the plates. As the capacitor has a fixed charge provided by the polarising voltage, variations in capacitance will cause voltage changes at the input of the high impedance source-follower buffer amplifier. Regular readers will remember how I accidentally made a condenser microphone when I was testing an audio amplifier in the February 2010 Homebrew.

The electret type of condenser microphone doesn't need a polarising supply across the capacitor plates because the capacitor is made from a special, permanently charged material called an electret. Modern electret microphones are small, light, inexpensive and capable of producing very high quality audio. Electret mics are used in mobile phones, camcorders, telephony headsets and anywhere else that a small high quality microphone is required. Electret microphones usually have a built-in FET (field effect transistor) buffer amplifier with a very high input impedance. The buffer amplifier

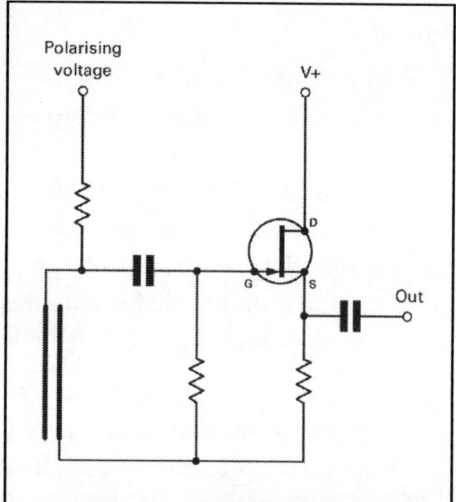

Fig 1: Basic condenser microphone. The left hand 'plate' is the diaphragm, which is moved in relation to the grounded plate by sound waves. This results in a voltage output that is impedance matched and buffered by the FET

requires a DC supply, typically in the range

1.5 to 12V at a current of less than 1mA.

DYNAMIC MICROPHONES

Fig 2 shows the configuration of a moving coil dynamic microphone. This type of microphone has a lightweight coil that is attached to the centre of the diaphragm. The coil is suspended in a magnetic field that is usually provided by a permanent magnet. Air pressure acting on the diaphragm causes the coil to move through the magnetic field, which induces a small electric current in the coil. The dynamic mic acts like a loudspeaker in reverse. Some small loudspeakers can make reasonably good microphones.

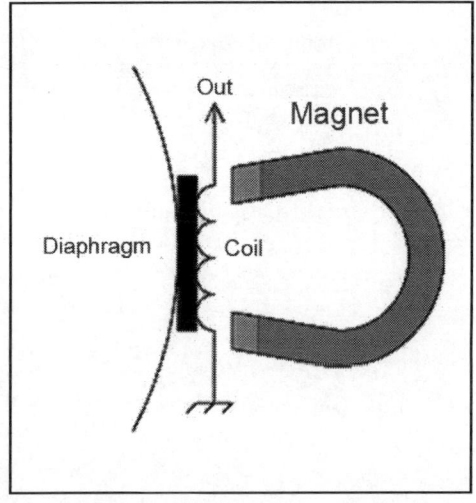

Fig 2: Basic principles of a dynamic microphone. A coil attached to the diaphragm moves in sympathy with sound waves and directly generates an electrical signal by induction.

Photo 1 shows a typical dynamic mic insert (left) and an electret condenser mic. The dynamic insert is a DM-9 type (Maplin QN39N), the electret is a type EM-60B (Maplin FS43W). SSB is not very demanding in terms of microphone frequency response. The bandwidth of an SSB transmission is usually less than 3kHz. A typical SSB exciter will require a reasonably flat response from about 200-300Hz at the LF end up to 3kHz at the HF end. Both of these microphones have a specified frequency response from 50Hz to more than 12kHz. Dynamic microphones have a specified load impedance, which is usually in the 200-600Ω range. The DM-9 specifies an optimum load impedance of 600Ω. It is usually acceptable to use a load impedance that is a bit higher than the specified value. Many microphone amplifier inputs have an impedance somewhere between 1kΩ and 50kΩ. Using a load impedance that is much lower than the specified optimum impedance may lead to low output, excessive distortion and poor frequency response.

Electret microphone elements require a very high load impedance, in the order of several megohms. Most if not all electret capsules (as in the Photo) have a built-in FET amplifier that meets this requirement. The optimum load impedance for the output of the FET amplifier will be a more reasonable value ranging from 1kΩ to about 10kΩ.

As most dynamic microphones don't have a built in amplifier, no DC supply will be required. Take care to ensure that DC is not accidentally applied across the termi-

Photo 2: The authors original desk microphone, which has given more than 20 years' good service.

nals of a dynamic mic. This could easily burn out the coil or damage the diaphragm. A coupling capacitor of about 10µF between the mic insert and the amplifier input will isolate the mic coil from the DC bias of the amplifier. Dynamic inserts with plastic bodies are usually symmetrical so that they can be used with either balanced or unbalanced amplifier inputs. Some metal-bodied inserts may have one end of the coil attached to the case ground. Electret inserts have the negative supply/mic ground terminal attached directly to the metal outer shell of the capsule.

There are two types of electret capsule in common use. One type has just two connection terminals, the other has three. **Fig 3** shows typical connections for both types, **Fig 4** shows the electrical configuration of both types. The three-terminal capsules (a) have a built-in load resistor for the FET amplifier. The DC supply is fed to one end of this resistor and the amplifier AF output is taken from the other end at the junction with the drain of the FET amplifier. The two-terminal capsules (b) don't have this resistor built in, so it is necessary to use an external resistor connected to the DC supply. The external resistor is typically about 2.2kΩ for a 5-12V supply and about 1kΩ for a 1.5V supply. The external coupling capacitor is typically about 1µF. A close inspection of the back of the capsule will show that it is very easy to identify the negative/gnd terminal because it is always connected directly to the metal case. Note that a three terminal electret can usually be used to replace a two terminal type by simply leaving the V+ terminal unconnected.

MICROPHONE AMPLIFIER

The output from a typical dynamic microphone is just a few millivolts (mV) when the operator is speaking directly into the mic and even less when the

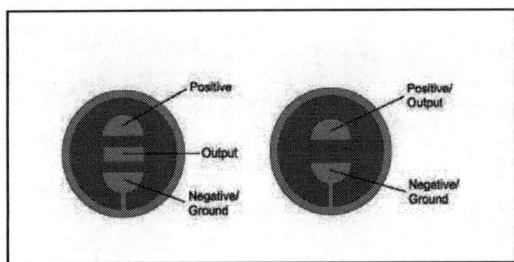

Fig 3: Typical electret mic connections: (left) three-terminal type (see Figure 4a) and (right) two-terminal type (see Figure 4b).

mouth is more than a few centimetres away from the diaphragm. Electret mics produce slightly more output: I measured about 20mV p-p at normal speaking levels. This may be enough to fully modulate some transmitters, but my homebrew rig and some commercially made rigs will require a higher level for proper operation.

The microphone amplifier in this month's construction project is based on my standard design as used in several previous projects [1]. I used the surface mount version of this amplifier as shown on p59 of the October 2009 RadCom. You can, if you prefer, use a DIL dual op-amp instead of the SMT components. The choice of op-amp is not too critical. I got good results using the LF353 dual op-amp. Even standard types like the humble LM4558 can be expected to perform well in this circuit. The schematic of this amplifier is shown in **Fig 5**. Some slight changes must be made to the circuit depending on whether it is to be used with the electret insert or the dynamic insert. For use with the electret insert, a 2k2 resistor was connected from the mic input to the DC supply rail. A 1µF capacitor was initially used for C1. The original R1/R2 values of 10k/100k were used. On-air testing of this configuration resulted in reasonably good reports on the audio quality. Some reports suggested that the audio was a little bit too 'bassy' compared to my usual microphone. Replacing C1 with a 100nF capacitor gives a slight bass cut at frequencies below 200Hz. On-air reports on this configuration were universally good.

For use with a dynamic microphone, the 2k2 resistor should be removed from the microphone amplifier input and C1 should be replaced by a capacitor with a much larger value: about 10µF works well with the DM-9 insert. If you use an electrolytic capacitor for C1, connect the positive side to the amplifier input and the negative side to the dynamic mic. To compensate for the low output of the dynamic mic, I changed R1 to 3k3 and replaced R2 with a 220kΩ pot so that I could set the amplifier gain to the required level.

DESK MICROPHONE

This construction project is a desk microphone that can be used with your home made or commercial transmitter. The design is based on my original desk microphone that has, so far, given more than 20 years of service, although I have had to replace the electret insert twice during this period. This microphone was built in an aluminium box, which gives good RF screening. The electret insert is mounted on a flexible metal 'gooseneck' removed from a cheap table lamp. A short length of plastic tubing was used to secure the electret mic on the end of the gooseneck. This microphone is shown in **Photo 2**. The Tx/Rx switch and mic gain

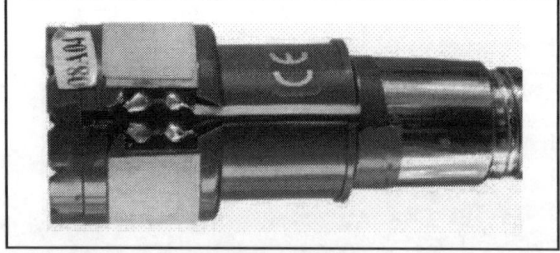

Photo 3: Detailed view of how the mic insert is mounted.

controls are on the top of the box. The small push button was used to operate a 1750Hz tone burst oscillator for VHF repeater operation.

A similar method of construction was used for the new microphone. An aluminium box (Maplin AB10 or similar) was used as the base. Unfortunately, a desk lamp with a suitable gooseneck was not available on this occasion. After checking out some of the local electrical and DIY stores, I found that the only desk lamps with a suitable gooseneck actually cost considerably more than a 'proper' microphone gooseneck. There are a few possible alternatives like the flexible map reading lights available from car accessory shops or flexible lamp goosenecks that are designed to plug directly into a USB socket. One very practical approach as used by EI2EUB is to use a length of co-ax cable (Westflex 103 is ideal) as the gooseneck with standard PL259 and SO239 connectors for mounting it on the base. I eventually bought a 6in gooseneck (Maplin YW72P), a DM-9 dynamic insert (Maplin QN39N) and, to make a really posh job of it, a mic windshield (Maplin LB35Q). The SMT microphone amplifier/LPF PCB described earlier was used to amplify the signal from the dynamic mic up to the 50-100mV level required by my transceiver. The PTT switch (although push-to-talk isn't

Photo 4: The finished microphone.

really an accurate description of this type of switch) is a standard DPDT single hole mounting type.

The DM-9 insert is a slightly loose fit on the end of a standard mic gooseneck. I used a thin layer of PVC tape to fill the gap. **Photo 3** shows close-up details of how the   microphone insert is mounted. It doesn't look very pretty, but the windshield does a good job of concealing the ugly bits. **Photo 4** shows the finished project.

POWER SUPPLY

The obvious choice of power supply for a microphone amplifier is a small 9V battery, but there are some disadvantages to this approach. If you do decide to use a battery, you will need to fit a switch so that the amplifier can be turned off when the mic is not in use. It would be possible to

use a spare pole on the PTT switch so that the amplifier is only powered up while you are transmitting. This approach can be problematicasitislikely to lead to audible pops and clicks on the transmitted signal. A separate power on/off switch is probably a better option, but if the switch is accidentally left on, the battery will run flat after a couple of days. You could fit a LED power indicator but, since it would draw more current than the mic amp, it might do more harm than good. In my experience, the best way to provide power for the microphone is to steal it from the radio. This is very easy to do if you use a home-made rig and you have a spare pin on the mic connector. Be careful about taking power from the mic socket of a commercially made rig. Many transceivers have a DC supply of about 8V available from the mic socket. It is possible to damage some transceivers by shorting this DC output to ground. Accidental short circuits can lead to blown voltage regulators and damaged wiring. You have been warned!

Fig 6 shows the configuration of the microphone, amp/LPF module and PTT switch. Note that I have included the mic gain pot shown in the Photographs.

I used a spare pin in the mic socket to provide power for the microphone amplifier. To eliminate the possibility of clicking/popping when changing from receive to transmit, the amplifier is always powered up, regardless of whetherIam transmitting or receiving. Even in these environmentally aware times, the 3-4mA drawn by the mic amplifier can be regarded as insignificant.

Iusedthe microphone for a few days without telling anyone that I had a new microphone. During this period, nobody noticed any change in my audio quality. Comparison tests have shown that the dynamic mic compares quite well with my old electret mic.

There is plenty of room in the base on the mic for additional circuits, such as

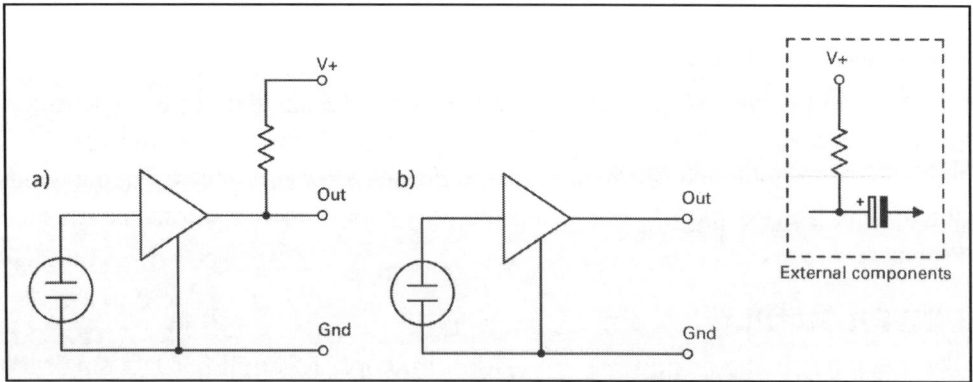

Fig 4: *Representative circuit diagrams of the (a) three - and (b) two-terminal electret mic types shown in Figure 3.*

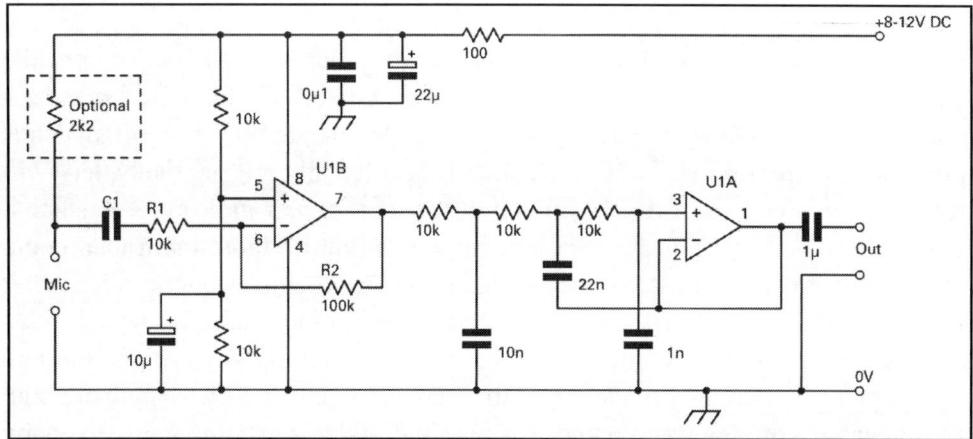

Fig 5: *The standard microphone preamplifier circuit. See text for C1, R1 and R2 values for electret and dynamic mics.*

oscillators for Morse code generation, two-tone testing, subaudible tone, DTMF generators etc... Other possible additions include level indicators, tone controls, equalisers, digital speech record/playback, AF or RF clippers/compressors – but please don't install a roger beep!

The microphone should be connected to the rig using screened cable that is grounded to the rig ground/chassis at one end and the metal case of the mic at the other. Special microphone cable has several wires with separate screening for the audio wire only. This guarantees good isolation between the audio wire and the other wires used for PTT and power. As this type of cable is not readily available, I use ordinary multi-core screened cable which has a single screen enclosing all wires in the cable. Old computer keyboard cables and some types of network cables are a good source of high quality screened wire. I have been chopping up old

AppleTalk RS-422 cables to make my microphone leads. RF signals on the PTT and/ or power wires could easily leak into the audio wire via capacitive coupling. To keep RF signals out of the microphone cable, it is a good idea to solder 10nF capacitors from the PTT and power pins directly to ground at the back of the microphone socket of a home made rig.

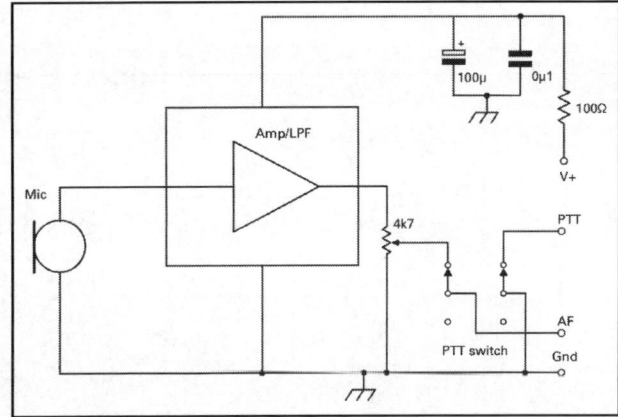

Fig 6: *Connection diagram for the microphone amplifier (Figure 5) and radio.*

REFERENCES:

[1] Homebrew August 2006, May 2009, October 2009.

FT-817 Remote Display
Simple design for safer mobile use

I was given my FT-817 about 6 years ago and have always loved the compact nature of this radio but I have never been able to mount it in a modern vehicle in such a way that I can read the display. The small LCD can be seen in daylight, just, but after dark with my ageing eyes it becomes absolutely impossible to read without pulling to the side of the road, removing my glasses and staring hard. To attempt this whilst in motion would have been suicidal.

I had a look on the web and found that Keith Dix, ZL1BQE had developed a hand held remote for a friend. He very kindly sent me a processor and EPROM and I assembled it into a working unit. Unfortunately the chip he used is no longer available and so I decided it would be a good learning experience to design and build a similar display module.

UPDATING THE DESIGN

I looked at doing this using a modern PIC chip such as the 16f628 and fairly early on decided that my assembly language skills were not up to the job. I think this puts a lot of people off having a play with these very useful devices. I decided I would program the chip using Basic and purchased Oshonsoft Basic, which comes bundled with a simulator package [1]. I wrote a few simple programs to flash LEDs and get a feel for the package then started on the FT-817 and the serial port. I wrote the routines in sections and actually did the two line display routines first. I then moved on to attempt to acquire 'real' data to display from the FT-817s serial port. Initially I was getting nowhere until by using the very useful serial port data display module in the simulator; I found the FT-817 was sending two spurious characters when the interface was turned on. I've not found out why they are generated but once

I read them and discarded that data we were off and running. The unit requires external power but it is usually quite easy to provide a low current 12V DC in a car. An alternative is to use an inline socket and plug to tee the power supply to the FT-817. The completed unit is shown in **Photo 1**.

Photo 1: Remote display installed on car air vent. The mount was created from a mobile phone holder like the one on the right (see text).

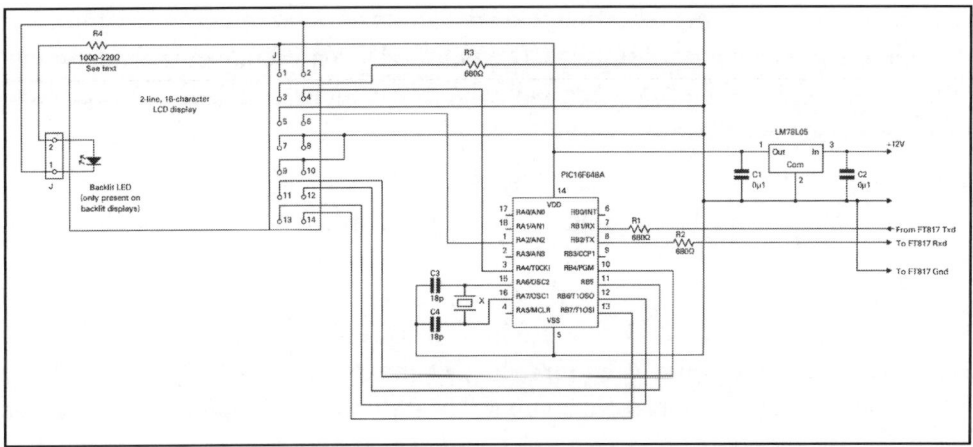

Fig 1: *Circuit diagram of the FT-817 remote display.*

My program outgrew the 16f628 so I moved it onto the 16f648, which has more program space. I used the Velleman Pic Programmer and experiment board available from Maplin; other development boards are available elsewhere and would have achieved the same result. The PIC HEX file can be downloaded from [2] and burned into the PIC with any suitable programmer.

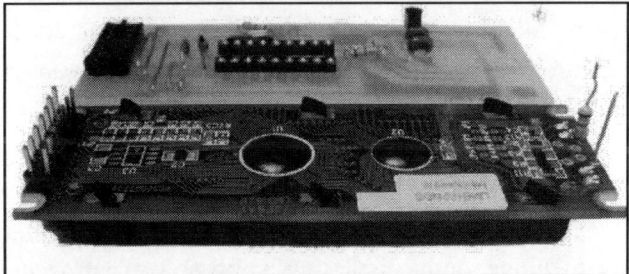

Photo 2: *General view of the display and controller PCB. Note pins, resistor and wire on LCD and modified socket on left of controller PCB.*

Photo 3: *Controller and display assembled together.*

HARDWARE CONSIDERATIONS

The hardware was designed around a standard 16 character x 2 line LCD module with 14 connecting pins on the short edge of the display. These are commonly available from a number of suppliers on eBay and elsewhere, as is the PIC. (Be careful; many 16 x 2 LCDs have a single row of pins on the long edge – check you're buying the right type!). The crystal is a 16MHz low profile type (normal crystals will work but may need to be laid flat on the PCB).

The full circuit diagram is shown in **Figure 1**. Note the very low component count! The circuit simply consists of the LCD connected to a minimumcomponent-count PIC driver circuit with a simple resistive interface out to the FT-817. All the clever stuff is done in software.

I designed the PCB (**Fig 2**) so that the display could be mounted on top of the board. **Fig 2** is reproduced at 100% size, and you can download the PCB CAD files from [2].

CONSTRUCTION

The component overlay is shown in **Fig 3**. Basic construction is straightforward, other than the connections between the LCD and processor PCB. I soldered 0.1" matrix pins salvaged from an old computer board into the data/power connector on the LCD. The pins plug into a 14 pin DIL socket on the processor PCB that has been modified to 0.2" spacing (instead of the usual 0.3") by cutting it in half lengthways and removing the remaining pieces of the centre frame. Solder the resulting socket strips onto the board so that what was the outer face of each socket now touches in the middle. I found the cheap sockets work best for this turned pin types are no use because they will not accept the large square connecting pins. Socketing the PIC will allow you to remove it to allow reprogramming if required.

If you are using a backlit LCD you will need to fit R4 and the ground link wire on the opposite end from the pins. I decided that soldering these in place was acceptable because the unit is not likely to be disassembled often; the resistor and wire form a useful hinge for the display assembly so you can access the PIC. R4 should be around

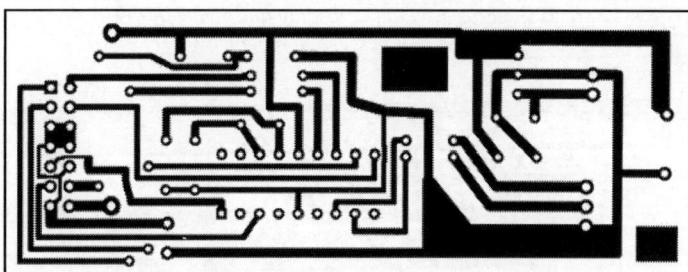

Fig 2: PCB foil pattern.

100-220Ω, lower values giving brighter backlighting. Check the unit's current consumption from the 12V supply with the backlight on; it must not be more than 50mA or the 78L05 voltage regulator will overheat.

Photo 2 shows the general arrangement of the populated PCB and display with the pins and sockets visible, and **Photo 3**

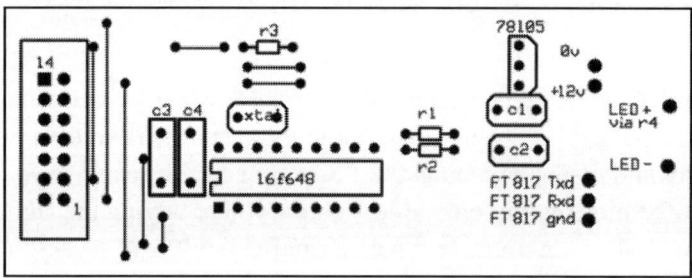

Fig 3: Component overlay. Note that there are 8 wire links.

shows how the two boards sandwich together for final assembly.

I leave the choice of case to you. I used a diecast box with a cutout for the display; you can use metal or plastic. The only important points to note are that you need an on/off switch and some way of getting power and data signals through the case.

I used a modified car phone holder to mount the completed box on the car vents. I obtained this from a supermarket for £2.99 and removed the phone jaws. I was then left with the back plate and the vent clip parts. The one I chose had two screws that held the two halves together so it was then easy to screw the back plate to the back of the display box. **Photo 1** shows an intact phone mount along-side the one used for the display. I'm sure you'll find a way to mount your display so you can see it easily!

IN USE

You must always switch on the FT-817 before the remote display. If you don't, the display just sits blank and will need its power cycled on and off before it will work. This is because of the spurious characters from the FT-817 mentioned earlier.

I found I was getting some interference from the unit at various spot frequencies in the bands. 2m and 70cm were clean, as were most of the HF bands apart from 10m and 6m. I found the best solution was to wind 10 turns of the data cable round a ferrite ring (Maplin 25mm diameter type) as close to the FT-817 end of the cable as possible. This filter then sits down under the radio between the seats – out of sight and out of mind. The unit has now had well over a year road test without any issues.

CONCLUSION

It has been suggested that this display will also work with the FT-857 and FT-897. Although I have not had a chance to test this, I believe this is probably the case. Maybe when I get some time I'll design a keypad section to go with the display. I'm sure other PIC Basic packages would work just as well if you want to experiment with your own code.

My thanks to Keith, ZL1BQE, for sparking off this project.

Andrew Britton, MM0MGB

WEBSEARCH

[1] Oshonsoft – *http://www.oshonsoft.com*

[2] PCB CAD and PIC files plus a longer version of this article with a detailed description of the BASIC source code are on the RadCom Plus website.

USB Digital Modes Interface
Professional grade interface at low cost

For several years I have been using a pair of transformer isolated 3.5mm audio leads and a simple one-transistor serial port keying arrangement for SSTV, PSK, RTTY and other soundcard data modes. The upgrade of the shack computer threw a spanner in the works because I now had only one serial port to share between my KAM TNC and the TX switch. The easy solution of course was a USB to serial adapter, which did the job well. However, I soon felt limited by having to change audio connections if I wanted to listen to audio from the PC, so a second sound card seemed the way to go. I thought about buying a commercial amateur radio rig interface but then reconsidered: I was sure I could make one myself!

An interface wish list was quickly turned into a list of requirements and how to fulfil them.

The important points were that the project should:

- be easily transferable between different computers, which really meant a USB connection
- have a built-in sound device, ie it would be an external sound card of some sort
- be transformer isolated

Photo 1: The finished interface.

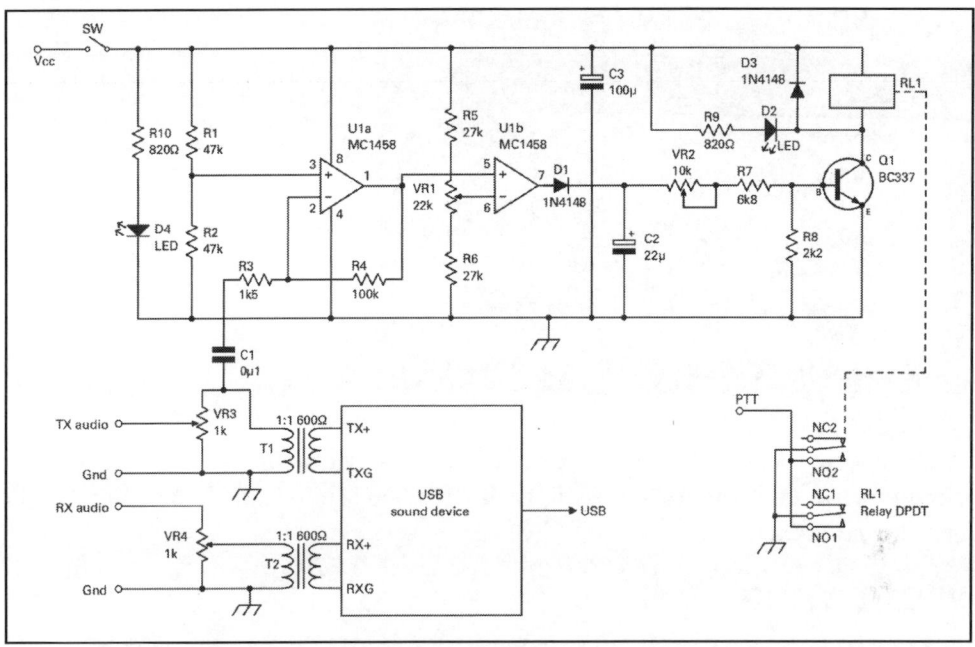

Fig 1: Circuit diagram of the USB sound card interface.

- be easy to set up, with accessible level and TX delay controls
- include an audio VOX circuit with an adjustable delay so there is no need for COM or LPT or radio VOX for the PTT.

GATHERING THE BITS

The first step was to find a suitable USB sound card. I was amazed to find memory stick style USB sound cards (**Photo 2**) on eBay for less than £3 including the postage from Hong Kong. Even if it proved totally unsuitable I decided it was a price worth paying in the pursuit of knowledge if nothing else!

A week later I was the proud owner of said device. A quick test was done by connecting the headphone output of the HF radio to the soundcard mic input, the new soundcard selected in the Digipan setup menu. QSOs then started to appear on the screen. The waterfall seemed a little noisy but that could be investigated later.

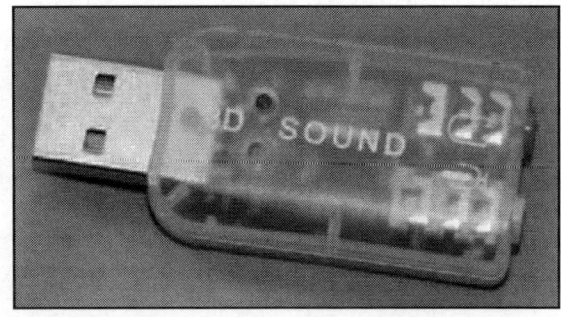

Photo 2: USB sound card.

Photo 3: Prototype built on stripboard.

A dip into the shack library of notes and circuits came up with an audio VOX circuit for which I can claim no originality. It's one I found somewhere and butchered for my own use.

PROTOTYPING

The first prototype was built on stripboard as a proof of concept project. This was the ideal way forward as it allowed several changes of component value and also changes of layout and circuit. As any visitor to my shack will tell you there are always many different projects around in various states of construction/repair/ redeign and this was no different. When I made my first QSO with the interface it was housed in a stylish enclosure -a padded envelope!

Happy in the knowledge of having made a QSO with the interface, it was boxed up – still in its stripboard incarnation -in a small case that I had handy (**Photo 3**). The small board tucked away at the left side of the unit is the unboxed USB soundcard. I removed its USB plug and replaced it with the lead from a pound shop USB mouse. Any USB lead could be used, including a USB extension lead if you don't fancy stripping the sound card down. The audio transformers are available from many of the regular component dealers, such as Rapid.

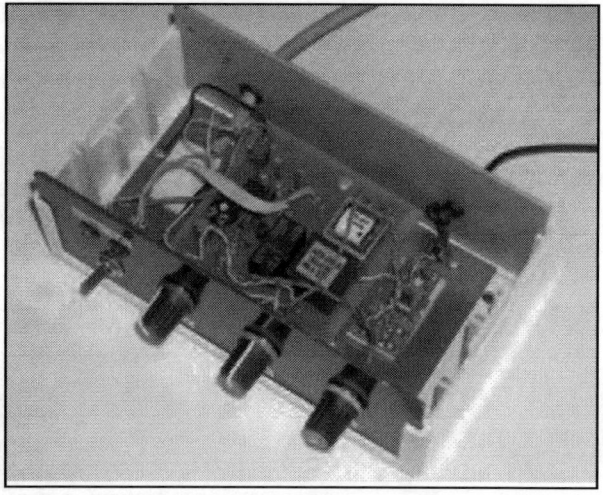

Photo 4: Inside the finished unit.

Electronics part no 88-2112, although I used ones salvaged from internal ISA modem cards that I bought for £1 each from a rally. (These also provided a supply of other valuable components including relays, opto-isolators and bridge rectifiers.)

The final circuit diagram is shown in **Figure 1**.

I now had a working USB interface, but as I often have to come up with ideas for my local radio club meetings [1], I decided to design a printed circuit board (Figures 2 and 3)sothat the design was easily reproducible. The board was produced using the inkjet photo paper, laser printer and iron method then etched and the final version was produced (**Photo 4**).

Since taking that photo I have installed a DIN socket in place of the flying lead to the radio. Different radios can be accommodated just by changing the lead. **Figure 4** shows the internal wiring of the unit, including the DIN socket. I also recommend fitting a series diode such as a 1N4001 and a 315mA picofuse in series with the +13.8V feed from the socket to the on/off switch.

IN USE

As mentioned earlier, when I use this interface the waterfall display is a little noisy (**Figure 5**).

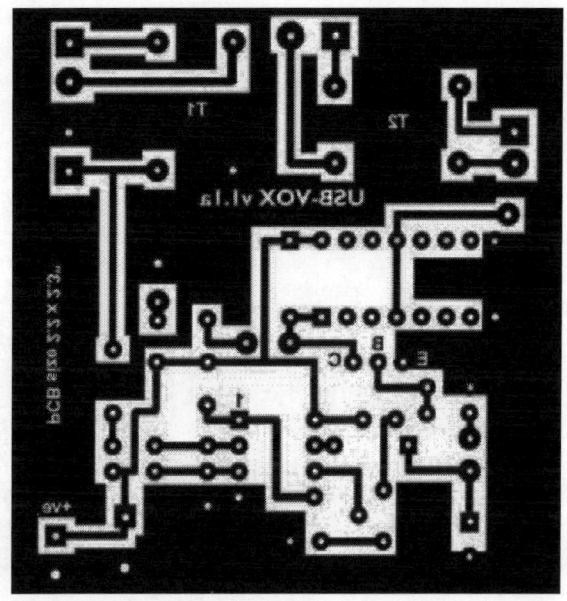

Fig 2: *PCB foil pattern of the USB sound card interface, reproduced at 1:1.*

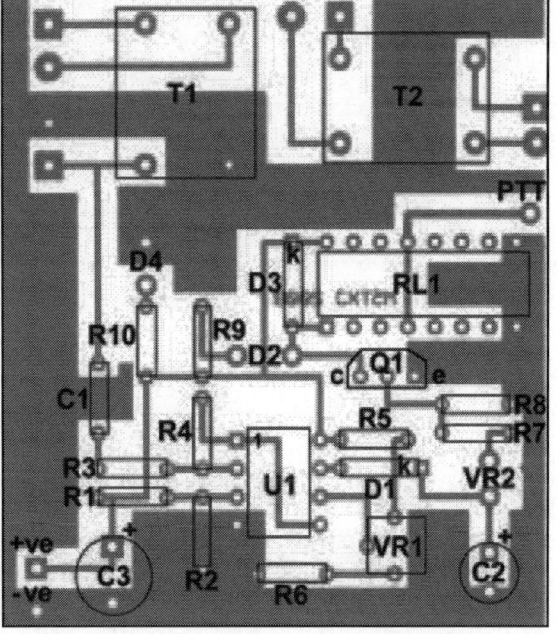

Fig 3: *Component overlay.*

In particular there are lines about every 1kHz across the spectrum. I've tried several different versions of the sound card (one advantage of their low price!) and found varying quality. This doesn't however seem to affect decoding ability. Following extended testing I've found that even using the 'worst' device I've got, I can still decode even the weakest of signals. If a really weak signal coincides with one of the 1kHz lines it can sometimes affect reception, but the very simple cure is just to re-tune the radio slightly.

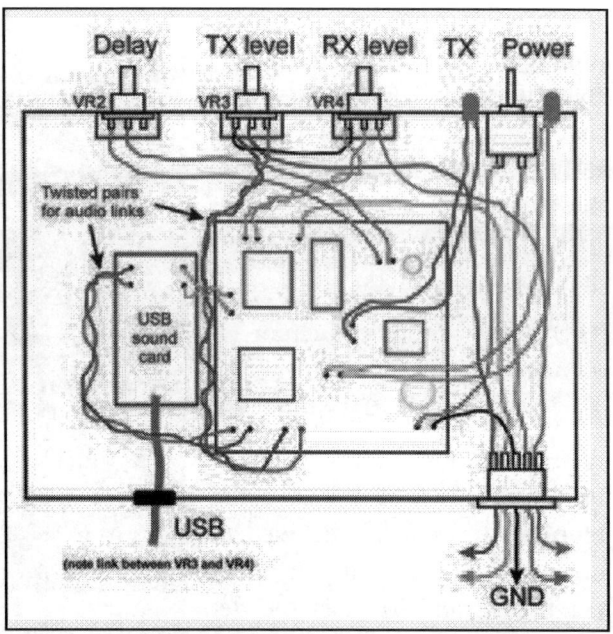

Fig 4: Internal wiring.

CONSTRUCTION

So how do you build your own? Layout is uncritical and you can use stripboard as I did for the prototype. It is good practice mount the transformers at 90° to each other. This may not be obvious from the photos as different models of transformer

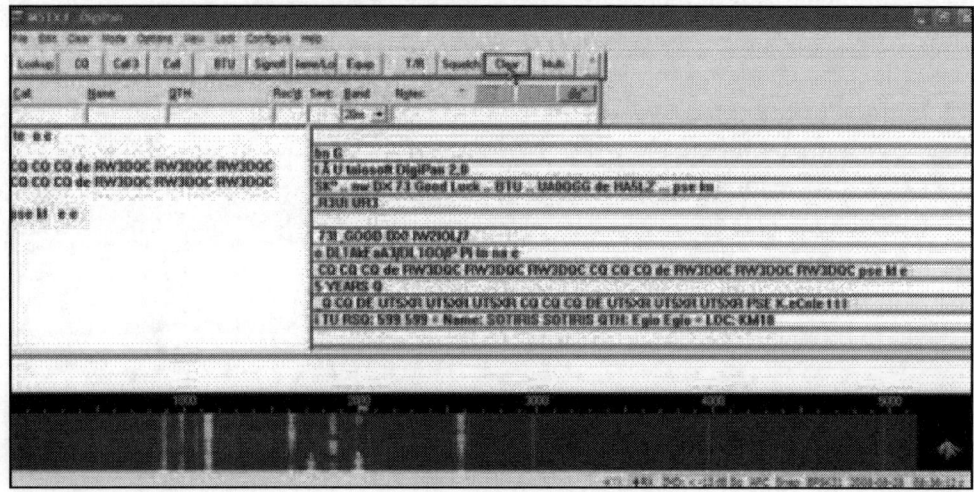

Fig 5: Typical DigiPan display using the 'worst' USB sound card.

were used as they were what were to hand. The other thing to watch is hum-inducing ground loops. If you decide to make a PCB using the design printed here then this aspect is mainly taken care of. My transceiver, an IC-735, has a 13.8V output on its accessory socket and I use this to power the interface. Since adding a DIN socket for the radio connection I have also added a standard 2.5mm DC input socket for use when the transceiver doesn't have this facility.

Setting up the interface is very simple and entails adjusting the preset potentiometer until the VOX circuit triggers the PTT in response to the audio output of the sound device. This is indicated by the red LED illuminating.

In use, set the RX audio level to give a clearly defined waterfall without excessive noise. As an aside, I find Digipan very tolerant of wide-ranging audio levels. On transmit, set the TX level to give sufficient drive to your transmitter whilst keeping ALC to a minimum (thereby improving your IMD performance). You will find that for PSK31 and similar modes you can set the delay control to minimum; other modes such as audio CW need the delay increasing. Of course if you do not need a USB interface, the board can be fed with audio direct from your computer's soundcard, possibly via a pair of 3.5mm sockets.

I've found this a very satisfying project, even as one of my simpler jobs, but must give many thanks to Chris, M0JKQ and Seamus, G7ITT for their assistance during testing of the interface.

David Shaw, M5TXJ/G8TXJ

COMPONENT LIST		C3	100µF 25V
R1, R2	47k	D1, D3	1N4148
R3	1k5	D2	Red LED
R4	100k	D4	Green LED
R5, R6	27k	Q1	BC337
R7	6k8	U1	MC1458 or TL072
R8	2k2		dual opamp
R9, R10	820R	RL1	DIL DPDT relay, 12V
VR1	22k preset		coil
VR2	10k lin	T1, T2	1:1 600Ω audio trans-
VR3, VR4	1k log		former, for example
C1	100n		Rapid 88-2112 (see
C2	22µF 25V		text)

REFERENCE

[1] Eden Valley Radio Society

– www.qsl.net/m5txj/evrs.htm

A 70cm UHF Handheld 4.12

Using The DRA818U Module ...built on perfboard

A homemade handheld 70cm amateur radio transceiver on a perfboard sounds like a challenge. With the SMD module DRA818U it looked feasible to me. All RF related functions are integrated in this module. The internet shows plenty of test circuits but never a complete transceiver including controller and antenna on a single board. When I found a small whip antenna with a SMB connector for 440MHz I started realizing this project on a perfboard.

INTRODUCTION

A homemade handheld 70cm amateur radio transceiver on a perfboard sounds like a challenge. With the SMD module DRA818U it looked feasible to me. All RF related functions are integrated in this module. The internet shows plenty of test circuits but never a complete transceiver including controller and antenna on a single board. When I found a small whip antenna with a SMB connector for 440MHz I started realizing this project on a perfboard.

Design goals. My goal was the design of a transceiver capable for direct or via repeater communication. Frequency and volume control in the fashion of commercial transceivers. A LCD for frequency and menu display and four keys for basic control.

Since the RF power of the DRA818U is only +30dBm (1W), an important part of the design is good antenna matching, especially for a handheld device. Designing and verifying the performance of a whip antenna mounted on a perfboard was one of the main goals.

The known problem of the module's poor harmonic suppression had to be solved. A low pass filter between the module and the antenna is mandatory in order to suppress the harmonics and to comply with the national law.

Photo 4: Inside the finished unit.

Components had to be found for PCB mounting. Power from a single rechargeable li-ion cell of 3.7V for the complete transceiver was a basic desire.

Photo 2: Side view of transceiver main board.

Photo 3: Underside of main board. Wiring is seen in Photo 9.

The DRA818U module. What is inside this compact transceiver module? The core is the RDA1846 from RDA microelectronic, providing the digital signal processing (DSP) for the TX and RX part. A microcontroller takes care of the RDA1846, the RF power amplifier, the audio process and squelching circuits. The UART provides the communication with the external controller.

All RF related functions for a complete transceiver are concentrated in this compact SMD module. I am using the 'U' type, standing for UHF-band. A 'V' version is also available that covers VHF. The manufacturer, DORJI Applied Technologies, specifies the frequency band from 400 to 470MHz. The dimensions are 36 by 19mm. Further technical data see [1]. Very comfortable is the supply voltage range of 3.3 to 4.5V. This matches the voltage of a single lithium ion cell perfectly. **Photo 1**, **Photo 2** and **Photo 3** give a goood impression of how the finished transceiver appears.

Software. A basic hardware and software design comprising a microcontroller, LCD, encoder and four buttons already existed from previous work. The supply voltage of these parts is 3.3V. These parts are of low current consumption and easy to supply from a single rechargeable lithium ion cell.

Specification		Software and control elements	
Operating frequency	430 – 440MHz	Key1	Toggle channel step 12.5 or 25kHz
RF output power	1W	Key2	Squelch in 8 steps
Modulation	Narrowband FM	Key3	Toggle 7.6MHz shift ON/OFF
	(NFM)	Key4	1750Hz tone (while PTT is pressed)
Frequency deviation	2.5kHz max	Pot1	Volume control
Suppression of		ENC	Rotary encoder for channel selection
harmonics	> 60dB	Key	PTT
Battery power	3.7V from one	LCD	2 lines with 16 characters for channel
	lithium ion cell		and menu

The software handling the keys, the LCD and the encoder had to be modified in order to support the functions of a transceiver. The encoder controls the frequency channel selection. The LCD of two lines x 16 characters shows the current frequency and other transceiver functions. The four keys control basic functions.

When rotating the encoder, the ENCODERA signal causes an interrupt to be seen at the INTA pin. In the interrupt routine the rotation direction is determined and a counter value is incremented or decremented. A flag indicates the main routine for update.

If one of the four keys is pressed, a pin change interrupt of the controller activates the corresponding interrupt routine in which the individual key is decoded. A flag indicates a service in the main routine. Every 100ms the main loop is initiated by a compare match interrupt of timer 1.

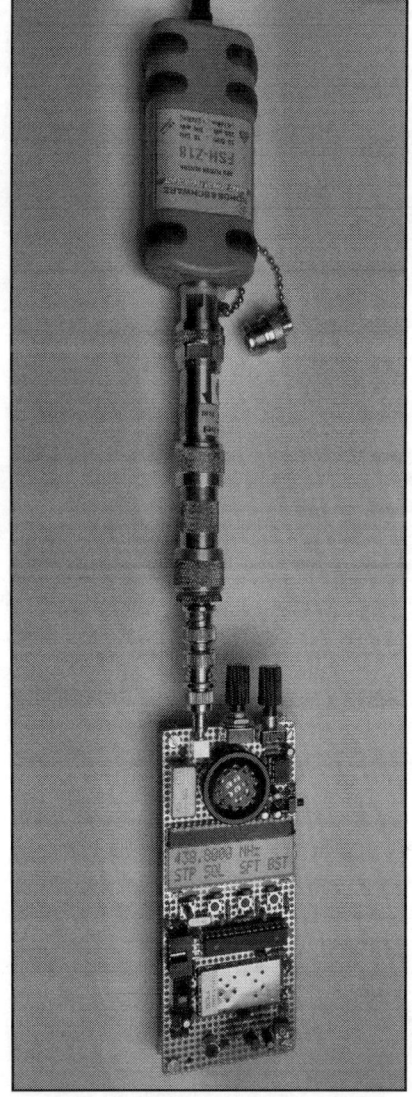

The output compare function of timer 0 is used to generate a 1750Hz burst signal for repeater access. CTCSS is supported by the module, but not implemented in the present version of my software.

Serial communication to the DRA818U. A simple serial data telegram is used to configur the DRA818U module. The default data format is 8 data bits, 1 stop bit, no parity and 9600kbps data rate. All commands in ASCII codes start with 'AT' and end with <CR><LF>.

The main command defines the following functions:

GBW	Channel space
TFV	Transmit frequency
RFV	Receive frequency
Tx_CTCSS	CTCSS transmit
SQ	Squelch
Rx_CTCSS	CTCSS receive

So a full command string is:
AT+DMOSETGROUP=GBW, TFV, RFV, Tx_CTCSS, SQ, Rx_CTCSS<CR><LF>

Details are described in the datasheet [1].

Photo 4: Measuring the output power.

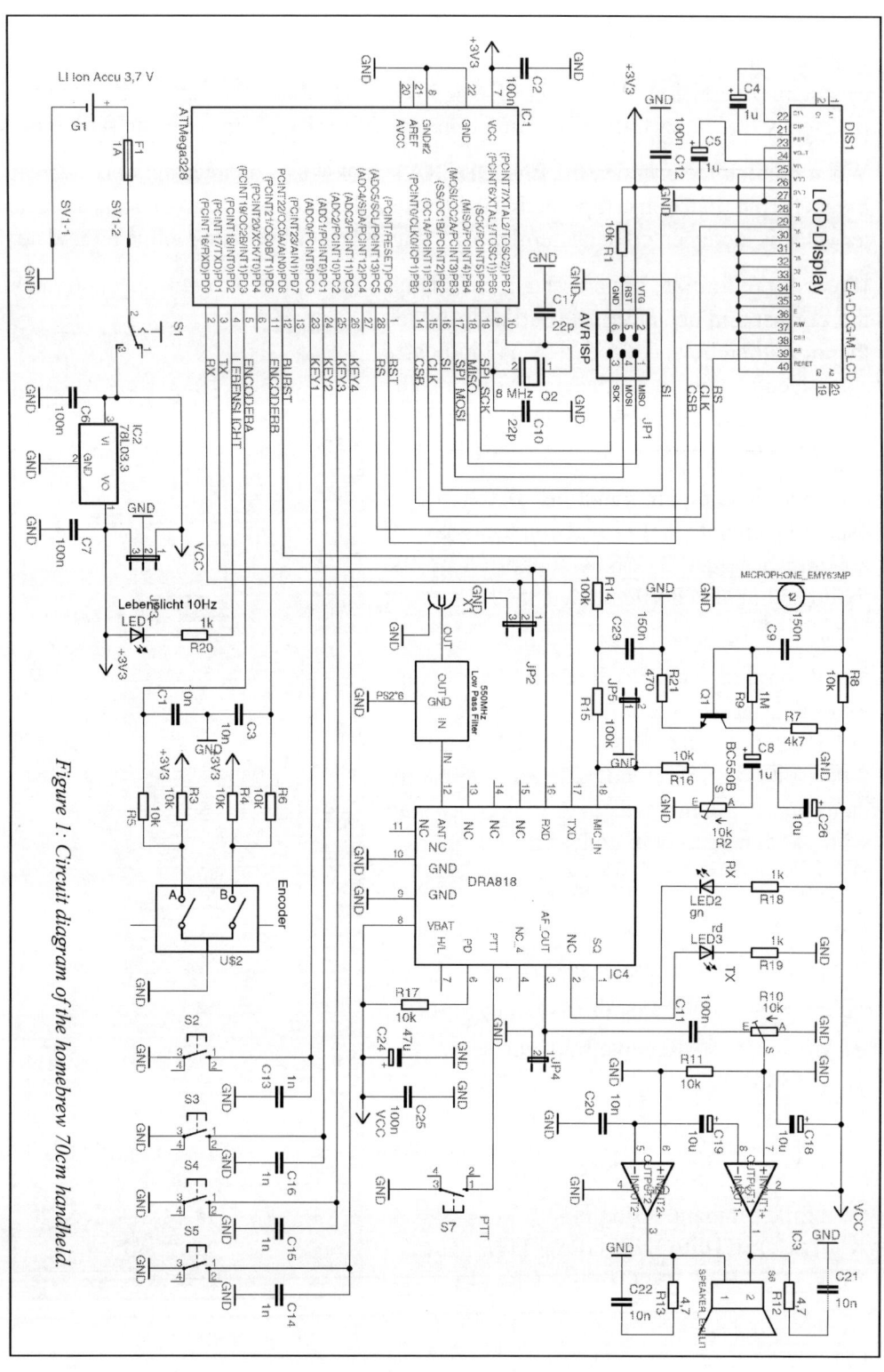

Figure 1: Circuit diagram of the homebrew 70cm handheld.

With each frequency change the full command is sent to the DRA818U.

The module responds with a string of data (see datasheet [1] but this response information is ignored by the software.

Photo 5: Test leads attached to the 550MHz low pass filter.

SCHEMATIC OF THE 70CM TRANSCEIVER

See **Figure 1** for the circuit diagram.

ATmega328P and periphery. The controller (IC1) is well known from Arduino projects. I choose the two line display EA-DOG-M (DIS1) because of the 3.3V supply voltage and its SPI interface. Only four pins of the controller are necessary for the data exchange. The encoder (U$2) and the four keys (S2~S4) built the digital part. The voltage regulator (IC2) produces a stable 3.3V from the lithium ion cell of about 3.5 to 4.2V. LED1 flashes while the controller is running.

DRA818U and periphery. Transistor Q1 amplifies the audio signal from the electret microphone. R2 adjusts the gain. R14 and C23 form a low pass filter for the 1750Hz burst tone. R15 reduces the amplitude to about 10mV at MIC_IN at the DRA818U module.

Between the module ANT pin and the antenna is the 550MHz low pass filter [2]. The RF output power was found to be 0.93W at 438MHz, measured at the board's SMB connector as shown in **Photo 4**.

LED2 (green) is lit if a signal is detected. LED3 (red) is lit while in Tx. S7 is the PTT.

Audio power amplifier. IC3 is an audio amplifier for low supply voltage. It comprises two channels operated in a bridge configuration. Output power is about 500mW into an 8Ω speaker. R10 is the volume control.

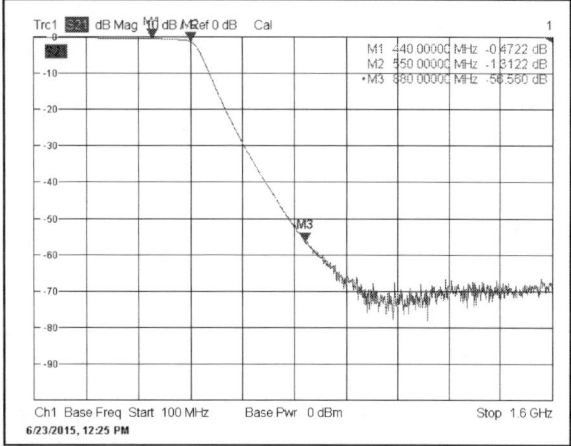

Fig 2: Screenshot from network analyser testing 550MHz low pass filter.

Frequency accuracy. At a default frequency of 438.800MHz I measured the transmitting frequency with a precision counter. The reading was 438.800300Hz. The offset of 300Hz in relation the bandwidth of 12.5kHz FM transmitter is negligible. The relative error is 0.68 ppm. Due to temperature increase while transmitting for about 10s I observed a drift of about 10Hz.

VERIFICATION OF THE LOW PASS FILTER

From prior transmitter measurements the requirement of a low pass filter became mandatory. I choose a Mini-Circuits PLP-550+ [2] available at [3].

With a network analyser I verified the insertion loss at the TX frequency and at the expected frequency of the 2nd harmonic. I soldered two coaxial cables with SMA connectors onto the leads of the filter. *See* Photo 5. **Figure 2** is a screenshot from the network analyser. I expected the isolation even higher when the filter is soldered onto a perfboard with a ground top layer.

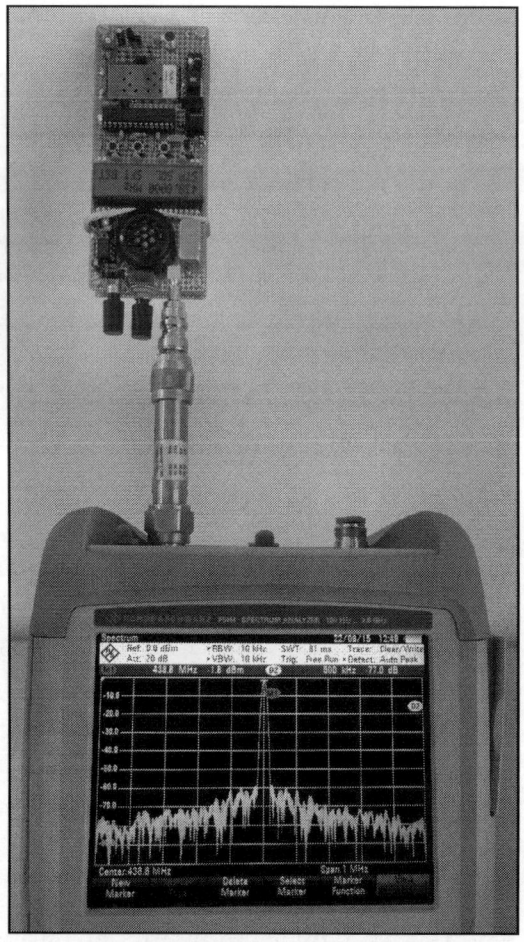

Photo 6: Test setup for checking harmonic suppression.

The insertion loss at 435MHz was measured as <-0.47dB, and at 880MHz it was -56dB. Third harmonic suppression at 1300MHz appeared better than I could measure.

TRANSMITTER HARMONICS

For verification of the harmonic suppression with the low pass filter in place I used the measurement setup shown in **Photo 6**.

The PTT was pressed continuously using a rubber band. A 30dB precision attenuator between the transmitter and the analyser was used to protect the input of the analyser.

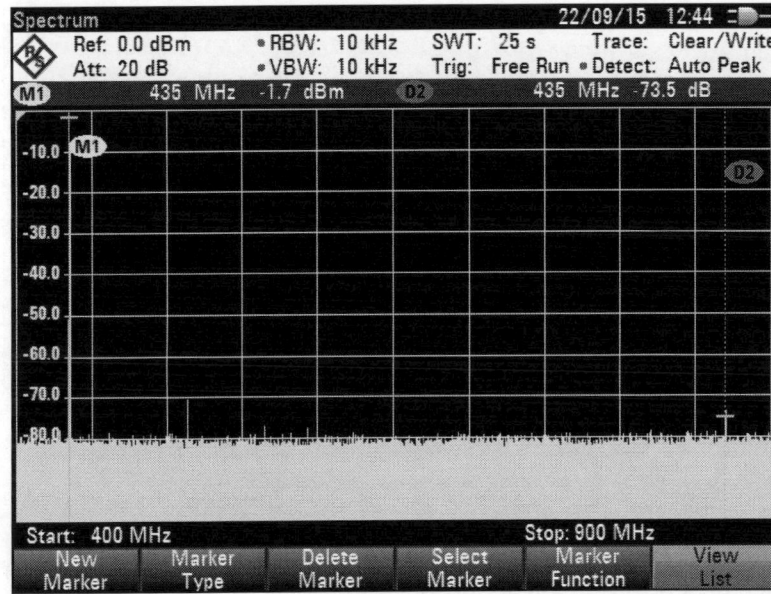

Fig 3: *Screenshot showing first and second harmonic measurement*

The first harmonic (fundamental) at 435.000MHz produced an indicated RF level of -1.7dBm (after the attenuator). The 2nd harmonic at 870MHz was found to have a suppression of -73.5dBc. The 3rd harmonic was not detectable. **Figure 3** shows the results screen from the analyser.

German Amateur Radio Law [4] regulates in § 16 section (4) the allowed emissions of harmonic. In the 70cm amateur radio band the specification is expressed as either total power of the harmonics or the suppression relative to the carrier. The measured result of -73dBc easily exceeds the legal requirement of -60dBc.

RECEIVER MEASUREMENTS

Audio frequency response. With a signal generator I produced a RF signal with 2.5kHz deviation, modulated from an external audio generator I injected this signal with an amplitude of -70dBm into the transceiver. With the probe of an oscilloscope on the AF_OUT pin 3 of the DRA818U I measured the frequency response, which is shown in **Figure 4**. I did not select any filters of the SETFIL-TER command.

The audio bandwidth is about 300 to 3200Hz. The DSP in the RDA1846 suppresses the audio below 300Hz for the CTCSS tones. The rising slope is remarkable. The dynamic range is probably better than in my measurement.

INTEGRATION OF THE ANTENNA

With a network analyser I verified matching of the 'rubber ducky' whip antenna. Therefore a SMB connector was positioned at the upper edge of a perfboard (**Photo 7**) where the position in the final design would be. The perfboard has a ground layer on the top which was used for the antenna. Photo 8 shows the antenna arrangement under test and shows the return loss with the board in a typical transmitting or receiving position for a handheld transceiver. This result was satisfying and no further matching improvements were necessary. The matching was found to be stable and the diagram did not change significantly with different positions of the hand.

Later with the dimensions of the final board the matching was checked again. In a range from 386 to 440MHz the return loss was better than -10dB, as shown in Figure 5.

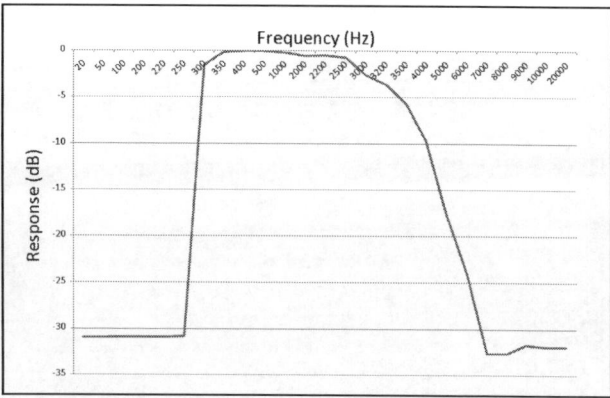

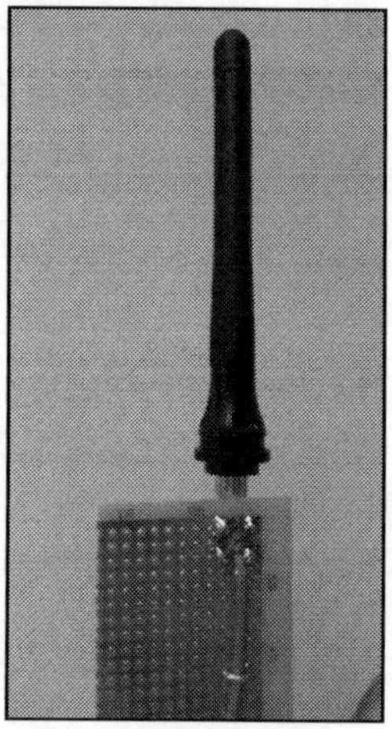

Fig 4: Receiver audio frequency response.

CONSTRUCTION ON PERFBOARD

Component positioning of the DRA818U module was difficult. Due to the different grid pattern of the perfboard and the footprint of the module a positioning onto the soldering side was impossible. The component side with the ground layer makes short circuits likely and a replacement very difficult. Mounting the module upside down onto the top layer known as "dead bug" gives good access to the pads underneath but is mechanically unsuitable for a handheld device.

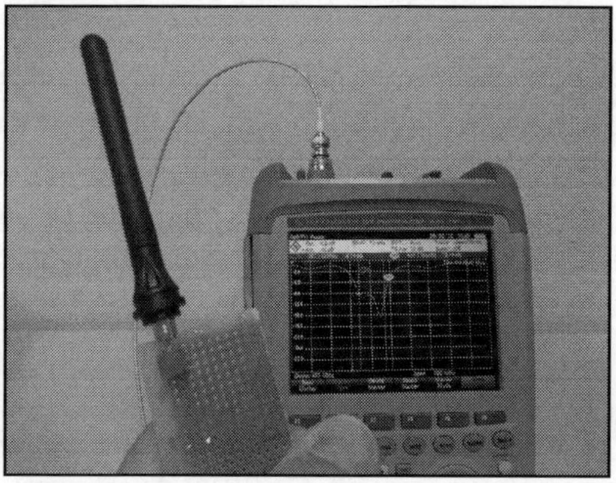

Photo 7: Testing the antenna matching.

Photo 8: Rubber duck antenna under test.

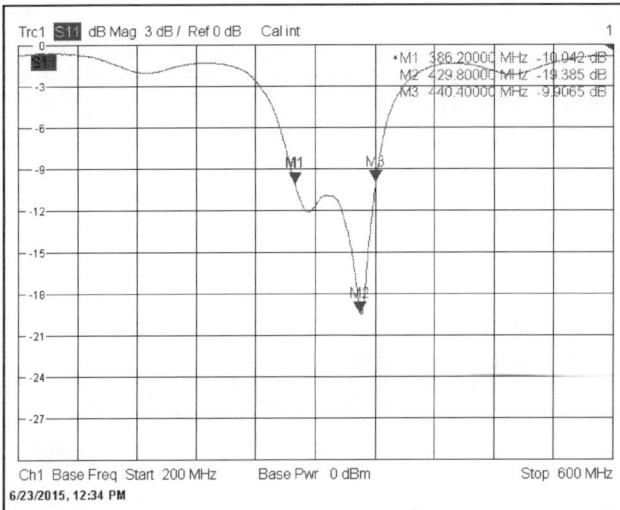

Fig 5: *Return loss plot for the rubber duck antenna.*

My solution was to saw a hole of the size of the module into the perfboad and to connect it to the solder side.

For the RF connection from the DRA818U to the filter and further on to the antenna I used a thin coaxial cable. *See* **Photo 9**. The dimension of the board was determined by the size of the LCD display and the amount of components to be placed. With four spacers I fastened a second board underneath. *See* **Photo 10**, the lithium ion cell fits well between.

CONCLUSION

Photo 9: *Solder side with wiring, showing coaial cable.*

Photo 10: *Side view showing li-ion cell.*

A perfboard with a ground top layer is essential for a proper matching and radiation of the antenna. The analogue and digital parts benefits from this layer.

The performance of this module is amazing, taking the physical dimensions into account. The RF power of 1W is well fed and matched to the whip antenna. This transceiver transmits and receives with narrow band FM with a 2.2kHz frequency deviation. For best audio the repeater or other station should use a similar frequency deviation (NFM).

From July 2016 l RadCom Plus

Clemens Verstappen, DL3ETW l *dl3etw@hotmail.com*

WEBSEARCH

[1] Dorji Applied Technologies:
www.dorji.com/pro/modules/Audio_voice_transceiver.html

[2] Mini-circuits Low Pass Filter PLP-550+

[3] FUNKAMATEUR-Leserservice: Majakowskiring 38, 13156 Berlin;
www.box73.de/product_info.php?products_id=3245

[4] Verordnung zum Gesetz über den Amateurfunk, §16 Technische und betrie-
bliche Rahmenbedingungen, www.gesetze-im-internet.de/afuv_2005/__16.html

Frequency Reference

Frequency is a measure of the number of events that occur in a given time period. Our local bus service departs every 20 minutes, or three times per hour. Less frequent events like the solar cycle will only occur once every several years. In the world of radio and electronics, frequency usually refers to the number of complete alternating current cycles in a period of one second, often happening millions of times a second.

Frequency is easily measured using a simple counter. To measure the high frequencies used in radio circuits, several digits will be required. A typical counter will display 6-8 digits. To accommodate the large numbers involved, a cascade of several counters is needed. Historically, the count would have been recorded on a chain of BCD (binary coded decimal) counter ICs. Most modern designs keep count in binary using a set of registers

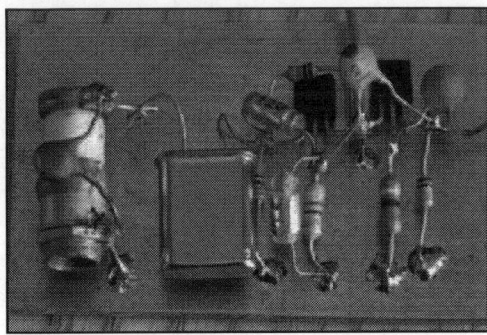

Photo 1: My prototype crystal oscillator.

(memory) in a microcontroller to store the count. The counter must be controlled or gated by an accurate timebase. *See* Homebrew for Jan-March 2008 for an in-depth look at frequency counters.

TIME

In the past, time was measured relative to the rotation of the earth, the orbit of the moon and the orbit of the earth around the sun. In the last few centuries, mechanical clocks have been the standard instrument for measuring time.

Today, the standard unit of time is the properties of the Caesium atom as "the duration of 9,192,631,770 periods of the radiation corresponding to the transition between the caesium-133 atom". For this reason, caesium clocks are regarded as the primary standard.

Photo 2: Connecting ground pins to the copper foil when building dead-bug style.

Accurate timekeeping is very important for science, navigation and for legal purposes. In several countries, accurate time and frequency standards are maintained by government organisations. Information from these standards is transmitted by radio. The radio-controlled clock in your kitchen is probably synchronised with the 60kHz MSF [1] signals from Cumbria or DCF77 near Frankfurt.

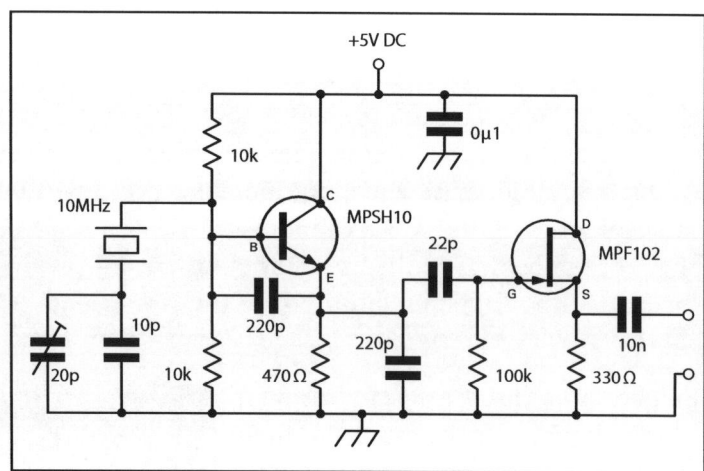

Fig 1: 10MHz crystal oscillator and buffer amplifier.

For the recent experiments with voltage references, I had to depend on a couple of local voltage standards and a couple of borrowed meters to check the accuracy of my measurements. Time and frequency calibration is much easier because very accurate time and frequency information is readily available from frequency reference was a 10MHz oscillator that was compared to a signal derived from the sync-pulses from a local UHF TV station. Since the change to digital TV, this reference is no longer available.

Currently available time and frequency standards include MSF (60kHz), DCF77 (77.5kHz), BBC long-wave (198kHz), RWM from Russia (4.996 and 9.996MHz) and WWV from the USA (2.5, 10, 15 and 20MHz).

The GPS network and other satellite navigation systems also provide a reference with excellent long-term stability.

10MHz or 9.996MHz Crystal Oscillator

Figure 1 shows the schematic of a crystal is calibrated against an off-air standard, it can be used as a short-to-medium term reference. After a few minutes of warm-up, stability is around 1 ppm (parts per million) or about 10Hz per hour. Even greater stability is possible if the unit is well insulated from sudden temperature changes. The oscillator uses an MPSH10 transistor in feedback is via a capacitive tap from emitter to in the common-drain mode (comparable to an emitter follower). The 5V DC supply is provided by a 7805 or similar voltage regulator. Power for all of this month's projects is provided by standard three-pin voltage regulators. To avoid unnecessary duplication, only one example will be shown, later.

The prototype circuit was built using pointto-point wiring on a strip of PCB laminate. If possible, the 220pF feedback capacitors and any capacitors in the crystal circuit should be high stability types. Polystyrene, silvered-mica or NPO ceramic types will be much more stable than standard ceramic capacitors.

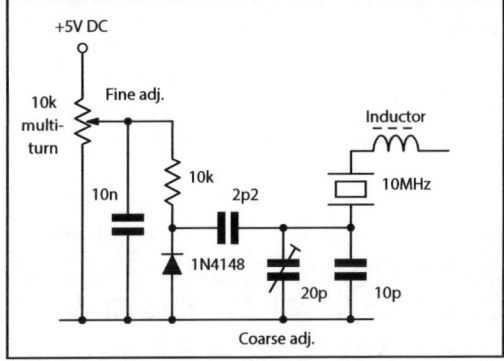

Fig 2: Modifications to enable fine tuing of a few Hz

TESTING

Oscillator frequency is adjusted by a 20pF trimmer capacitor. The values shown allowed tuning to precisely 10.000MHz using my current batch of 10MHz crystals. If your crystals have different characteristics, it may be necessary to capacitor. Frequency will be inversely proportional to capacitance. I set the oscillator to 10MHz using my bench frequency counter and then checked the value against off-air 10MHz time signals and my GPS disciplined, oven-controlled crystal oscillator (OCXO) (March 2008). The unit was then powered down and left for one hour.   to within 1Hz/hour for several hours.

The use of a standard trimmer capacitor a multi-turn, piston trimmer for tuning. This on the relatively weak time and frequency signals on 10MHz.

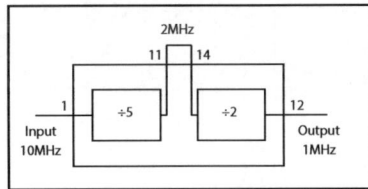

Fig 3: Simplified block diagram of how to set up the 7490 as a divide-by-10 counter.

I find the signal from RWM on 9.996MHz is a better reference. There is usually a lot less interference on this   usual standard. 9.996MHz crystals are not so readily available, but it will often be possible to 'pull' a 10MHz crystal down to the required frequency by placing a small inductance in series with the crystal.

Figure 2 to the oscillator. An inductor is placed in series with the crystal. The inductance value is very small compared to the motional inductance of the crystal. A value in the low μH range should be adequate. 20 turns on a T50-2 toroid is a inductance may have an adverse impact on stability or might even prevent oscillation. If you can't achieve the required frequency, try using a different crystal. Some crystals will shift more easily than others, even when they are from the same manufacturing batch. Figure   or minus a few Hz. The assembled oscillator is shown in **Photo 1**.

FREQUENCY DIVIDERS

A frequency divider will allow an oscillator to be used at sub-multiples of its frequency of oscillation. Dividers based on digital and logic devices like counter ICs will produce a square-wave output that is rich in harmonics. Such a signal can be very useful for calibrating receivers or test equipment. The most common types of counter are binary and BCD (binary coded decimal). The binary types give a division ratio and so on. The decade types are particularly useful for division ratios that are multiples of ten – 10, 100, 1000…

My usual first choice for a simple decade counter is the 74LS390 and similar devices from other logic famillies (74HC etc). The 390 is a dual counter, which will save on space and cost when several stages are needed. As   has a divide by 5 stage and a divide by two stage. This provides a very easy way to achieve division ratios of 2, 5 or 10. **Figure 3** shows a to divide by 10. This particular arrangement gives a symmetrical square-wave output with

A 10MHz oscillator as previously described can be used with a chain of dividers to produce multiple outputs at lower frequencies. This makes a very useful 'frequency marker' for testing and calibrating receivers or test equipment such as spectrum analysers. **Figure 4** shows a divider

If required, the circuit can be extended to produce additional outputs. Adding another decade counter would produce a 10kHz output from a 10MHz in-

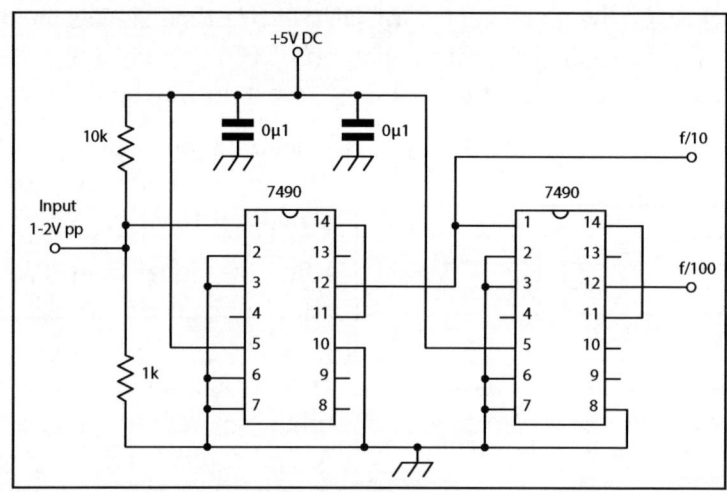

Fig 4: Practical divide by 10 and 100 circuit.

put. Using the divide by 2 stage alone would allow outputs at 5MHz, 500kHz or 50kHz. The outputs can be taken to separate output sockets or to a switch so that the selected output can be taken to a single output socket. Note that the outputs are completely unbuffered. If this unit is to be used as a frequency marker, it would be a good idea to place capacitors and resistors in series with the outputs, as shown in **Figure 5**. The capacitors will protect against accidental application of DC to the output terminals. The resistors will limit current in and out of the pins.

Higher values would offer greater protection, but would result in significantly lower output.

TESTING

The simple crystal oscillator was connected to the input of the first divider. The 5V supply for the oscillator/buffer and the divider was provided by a 1A voltage regulator. The unit was powered up and I noticed the ICs and the regulator were running a bit warmer than expected.

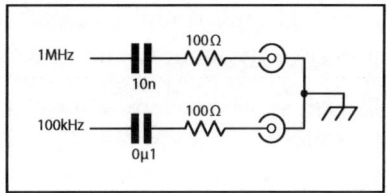

Fig 5: A capacitor and resistor between the divider outputs and front panel sockets affords a measure of protection.

A quick check with a voltmeter revealed a supply voltage of 8V. I'd managed to install a 7808 instead of the intended 7805!

As this is well above the absolute maximum voltage limit for the 5V logic ICs, the unit was powered down immediately and the regulator replaced with the correct item. Happily, the ICs seem to be undamaged and the outputs produce nice clean square waves at 1MHz and 100kHz. The harmonics are easily heard at HF and at least up to 144MHz.

OCXO

Figure 6 shows a 10MHz oscillator based on a packaged oven controlled crystal oscillator (OCXO). I used a CQE unit in this project. I have also seen similar units under the Temex and Isotemp brands. Surplus OCXOs are easily found on eBay and occasionally at rallies.

The 12V supply voltage and package pinout shown is correct for this device.

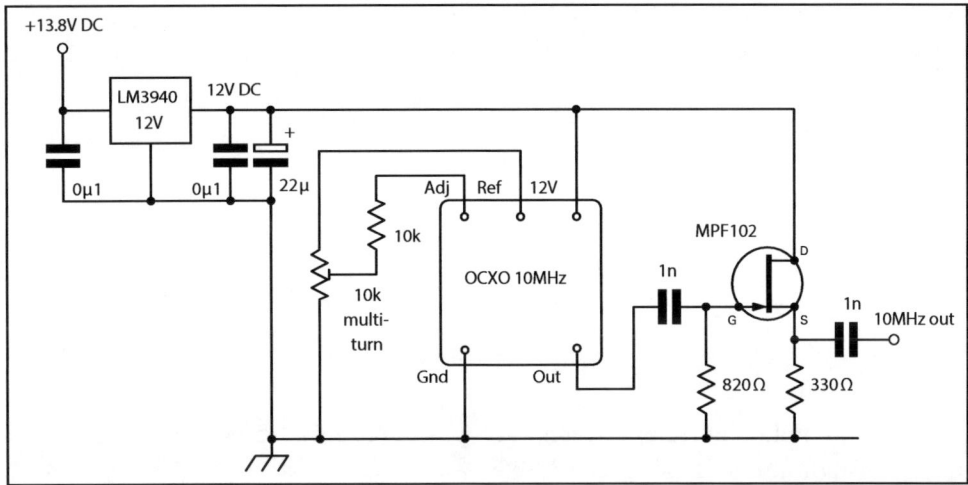

Fig 6: OCXO circuit.

Don't assume it will also be correct for yours. Check carefully before applying power to an unknown unit.

Power is supplied by a 12V, 1A regulator. A low-dropout (LVD) type like the LM3940 is recommended in cases where the unit will be powered by a lead-acid battery or a 12.5-13.8VDC bench supply. This type of oscillator offers stability measured in parts per billion. A typical spec would be ±5ppb per day over the full temperature range of the device and typically better than 1ppb per day at normal room temperature. The output frequency can be trimmed by a couple of Hz using a multi-turn cermet pot. The output buffer is a JFET source-follower. The circuit was built on Veroboard and the finished unit is shown in **Photo 3**.

GPS disciplined oscillators

The trouble with having such an extremely stable oscillator is that you need a very accurate standard to calibrate it against. I use a Navsync CW12-TIM timing GPS as my long-term frequency stability reference. The 10MHz output from this receiver has extremely good long stability, comparable to the atomic clocks on the GPS satellites, when averaged over a fairly long period.

There is no direct connection between the GPS receiver and the OCXO. I use a dual beam oscilloscope to compare the phase of the two 10MHz output signals: one from the GPS receiver, the other from the OCXO. Using this method, it is quite easy to adjust the OCXO to within one cycle of drift in several minutes. This corresponds to a frequency error of one part in several billion – more than good enough for my purposes. The CW12-TIM receiver is shown in **Photo 4**. The receiver requires an external GPS aerial (magnetic mounting mobile types are ideal) and a 3.3VDC supply. As usual, I used a 3.3V 1A regulator. The GPS unit produces output in the standard NMEA data format. It also has a very accurate 1 pulse per second output.

Eamon Skelton, EI9GQ
hbradio@eircom.net

Websearch

[1]   www.npl.co.uk/science-technology/timefrequency/products-and-services/time/msf-radiotime-signal

A Colourful Voltage Monitor

Probably the most common unit used in electrical, electronic and radio engineering is the volt, which is a measure of the amount of energy associated with the electrical charge of electrons at a particular point. One volt is defined as one joule of energy per coulomb of charge, and there are two main situations where it is encountered. The first is where a voltage is 'dropped', for example across a resistor, and this is a measure of the energy lost as a current flows (ie electrons move) against that resistance. This is usually referred to as potential difference (PD). The other is where a device 'pushes' the electrons, in the form of an electrical current, around a circuit, as does a signal generator or a battery. In this case the voltage is usually referred to as the electro-motive force (EMF). So when measuring the voltage of a battery we are measuring how much energy the battery is capable of giving to the circuit of which it is part. As the energy stored in the battery is used up, its output voltage falls, an indication of the falling amount of energy available to 'drive' the electrons. Often the degree to which a battery has been 'used up' is of more interest than the actual voltage output, and in this case an indication of a voltage 'range' is more useful.

WHAT IT DOES

This project uses a series of LEDs to indicate when a voltage falls below 12V, when it is inside an acceptable range of 12 to 13V, when it is inside a second acceptable range of 13 to 14V, when it is a little too high at between 14 and 15V, and to warn when it is above a maximum acceptable 15V. With these ranges, it is ideal for monitoring car battery voltage, and when using rechargeable batteries to power equipment around the shack or 'in the field'.

HOW IT WORKS

To give a meaningful output in the form of a sequence of ranges, a series of reference points is needed with which to compare the test voltage. These reference voltages can easily be generated using a resistance divider chain. However, the main reference from which all these are derived needs to be stable at all times, and it is not sufficient simply to connect the resistor chain across the supply battery. If this technique were to be used the reference voltage would of course change as the supply battery aged or when it was subject to different load currents. In this project a Zener diode is used to achieve a stable reference voltage. These devices are semiconductor diodes connected 'backwards', ie reverse-biased and, when used like

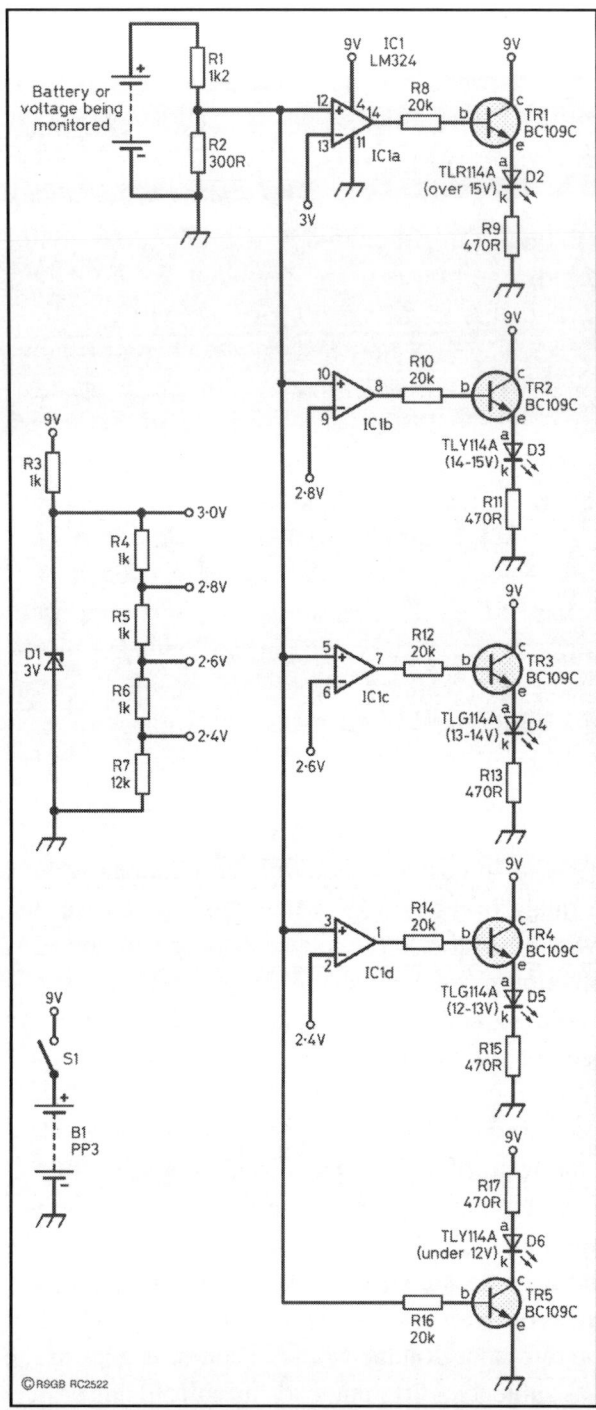

Fig 1*. The device works by comparing a divided fixed voltage to a divided sample of the voltage being monitored.*

this, the normal flow of current through them is blocked. However, all semiconductor devices have a 'leakage' current which is a small current that passes in the 'wrong' direction, and Zener diodes are designed to give a particular fixed voltage across them when this leakage current is flowing.

Each reference voltage from the resistance divider chain is fed to the inverting input of its own comparator, the test voltage being connected to all of the non-inverting inputs. Thus the output of an individual comparator will be around 0V unless the test voltage rises above the comparator's reference voltage, at which point it will switch over to a value close to the positive supply voltage. By connecting each output to an LED with its own driver transistor to provide enough current, the LEDs will light up in turn as the test voltage increases.

Since the test voltage will normally be higher than the circuit supply voltage, the test voltage and the reference voltages are both scaled down by the same amount, in this case by a factor of five. The test voltage is scaled by a simple voltage divider, the reference voltages by using a 3V Zener diode (the maximum reference voltage before scaling being 15V).

The '<12V' indication is given by switching the LED driver directly from the scaled test voltage. In this way a false '<12V' indication is not given when there is no test voltage connected.

CIRCUIT

The circuit diagram for the voltage monitor is shown in **Fig 1**. The comparator functions are obtained by using the operational amplifiers (op-amps) of IC1 wired without any feedback. In this configuration the op-amp's high gain means the output can only ever be equal to the positive or negative (0V) supply voltage, depending upon whether the voltage at the non-inverting input is more positive or more negative than the voltage at the inverting input.

The transistors all operate as switches, ie they are either 'on' (conducting) or 'off' (non-conducting). When they are 'on', the transistor passes enough current to light the LED (the op-amp output alone cannot reliably provide enough current to do this). The values of resistors chosen in the transistor / LED part of the circuit are those that limit voltages and current flows to the working values of the transistors and LEDs.

CONSTRUCTION

The stripboard layout for the project is shown in **Fig 2**. The LM324 quad op-amp integrated circuit, the transistors and the LEDs all need to be connected the 'right way round', and **Fig 3** shows the correct orientation of these devices. An important point to note is that the 'positive' test lead needs to be connected to the more positive terminal of the battery under test, so it is useful to use red and black wires respectively for the test leads.

COMPONENTS

An LM324 quad op-amp was chosen as it gives four devices in a single package and can be powered from a single

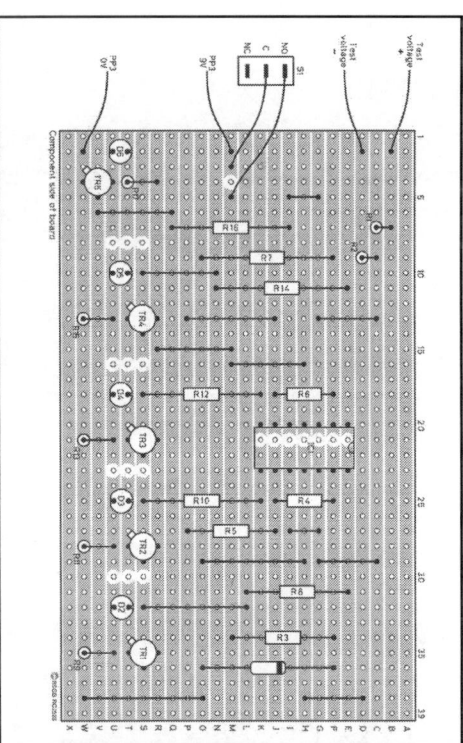

Fig 2. Stripboard layout of the voltage monitor.

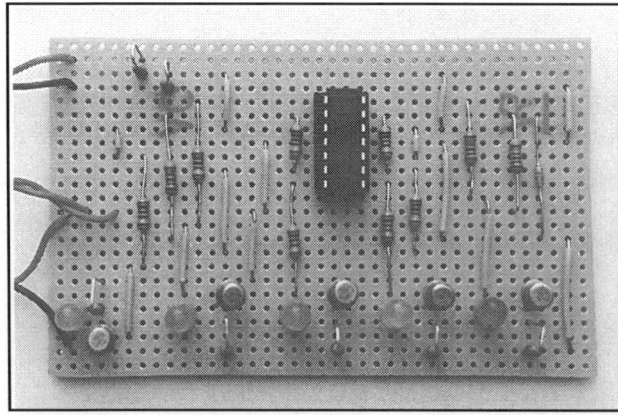

Fig 3. Semiconductor connections.

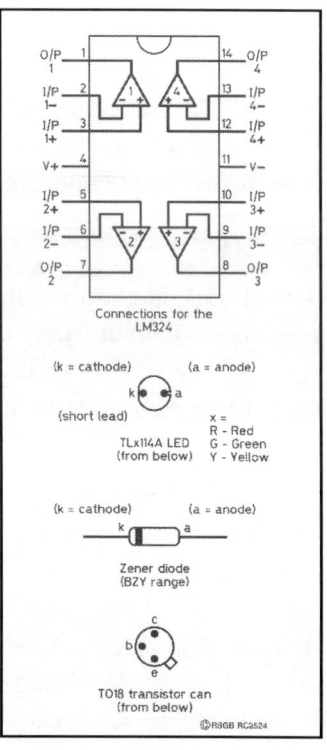

supply voltage. The transistors are a general-purpose npn type, and similarly the LEDs are different colours of a general-purpose type. The lower the tolerance value of the resistors used for the voltage divider chain, the more accurate are the reference voltages.

IN USE

Simply connect the test leads to the voltage being monitored (ensuring that the 'positive' test lead is connected to the more positive terminal), and switch on!

The completed unit. The components are laid out on stripboard. The monitor could be put into a project box or the LEDs mounted separately if desired.

PARTS LIST

Resistors - All resistors metal film, 0.6W 1%
R1 1.2k
R2 300
R3 – 6 1k
R7 12k
R8, 10, 12, 14, 16   20k
R9, 11, 13, 15, 17   470

Semiconductors
IC1   LM324
TR1 – 5   BC109C

D1 BZYC3 Zener, 3V 0.5W
D2 TLR114A red LED
D3, 6 TLY114A yellow LED
D4, 5 TLG114A green LED

Additional items
S1 SPDT
Stripboard
PP3 battery and clip
Crocodile clips

An Amplified RF Probe

When constructing radio-related projects, an RF probe is an item of equipment which is very useful to have around. It is also something simple to build. This RF probe employs a field-effect transistor (FET) amplifier to increase its sensitivity.

BACKGROUND

RF probes are often built to be used in conjunction with a multimeter. This ties up the multimeter, which cannot then be used to make other measurements at the same time. This simple project, the circuit of which is shown in **Fig 1**, adds an FET amplifier to the basic probe. The whole project is housed in a small metal box, complete with 9V battery.

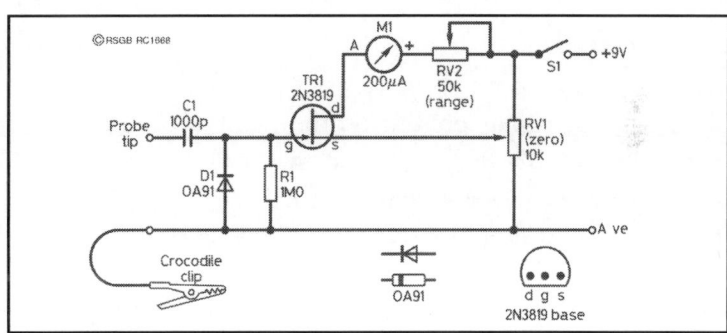

Fig 1. The amplified RF probe uses D1 to rectify an RF signal, TR1 to amplify it, then M1 to display it.

The meter required is not critical in nature. The prototype used a scrap item from a tape deck, and had a full-scale deflection (FSD) of 200µA.

CONSTRUCTION

Components are wired up point-to-point, as shown in **Fig 2**. The probe is soldered to the tag strip, the insulation on it keeping it clear of the metal case. The on-off switch, S1, is part of a switched potentiometer (RV1), but you could just as easily use a toggle or other type of switch.

The probe itself is made from stiff copper wire. This needs to be cov-

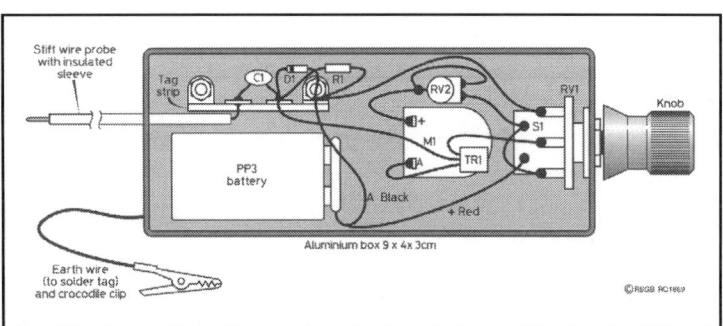

Fig 2. Physical layout. Note the insulation on the probe where it passes through the metal box.

ered with sleeving along most of its length, especially where it passes through the hole in the box. Diode D1 can be any small, germanium, point-contact diode, but a Schottky type would make the unit even more sensitive.

The photograph shows the connection of the short earth lead to a solder tag on the metal case. The crocodile clip on the end of this lead must be connected to the earth of the equipment being tested.

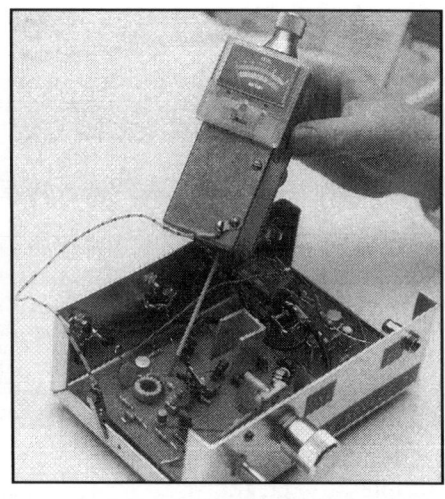

OPERATION

Switch the probe on, then adjust RV1 so that, with no signal present on the input, the needle of M1 is just above the zero mark. With the probe touching an RF sig-

In use, attach the crocodile clip to the ground (earth) of the equipment you are checking.

nal source, you should see the meter needle rise. RV2 is used to control the sensitivity of it. Although it would be possible to calibrate the RF probe against a known millivoltmeter, in most instances it is only required to either see whether a signal is present or not, or to adjust a circuit to peak a signal.

An Op-amp Tester

When building a circuit, it is not unusual to find that it doesn't work first time. After fault finding it's not unusual either to find that one of the supply rails had been connected to either an input or output of an operational amplifier (op-amp). This could be either because of an incorrect link, an unnoticed short-circuit between copper tracks, or a direct connection between the IC pins due to a 'whisker' of solder. With a stripboard project it could also be due to an intended break in one strip that has not been completely cut through, or a burr of copper that is shorting to an adjacent track. It might also be due to a required break that has been omitted.

When the fault has been rectified and the circuit still doesn't work, it then isn't possible to say whether this is now due to another fault present or whether it is because the op-amp was damaged by the initial fault. The best solution to this dilemma is to check independently that the op-amp is working correctly. This simple project will perform just such a check. It can also be used to check that an op-amp salvaged from unwanted equipment for use in another project works correctly. It will prevent time being wasted in checking for construction faults, when it is in fact the component that is faulty.

HOW IT WORKS

When a component fails, particularly a semiconductor device, it usually fails catastrophically, ie it fails completely and doesn't just work half-heartedly.

In the case of an op-amp, this usually means that the output goes to the value of one supply rail or other and stays there. Another possibility is that the output goes to some fixed DC value between the two supply rail limits.

This circuit works by incorporating the op-amp under test into an astable oscillator circuit. If the circuit oscillates, the op-amp is fine; if not, it is damaged. In order that an oscilloscope is not needed to observe the output waveform, the op-amp output is connected to a detector circuit and the output from this is connected to an indicator section. The indicator section drives two LEDs: a green one to indicate that oscillation is present, ie the op-amp passes the test; and a red one to indicate no oscillation, ie the op-amp fails the test.

THE CIRCUIT

Components R1 to R3 and C1, in combination with the op-amp under test, form an astable oscillator (see **Fig 1**). When this is operating correctly, the output Vtest is a square wave with a frequency of about 1kHz. Otherwise Vtest will be a DC voltage.

If Vtest is a DC voltage it will be blocked by C2 and Vdiode will fall to 0V as the right-hand plate of C2 discharges via R4 (with time constant C2 × R4). D1 then isolates IC1 from this part of the circuit and both inputs of IC1 are connected via resistors to 0V. The non-inverting ('+') input, however, is connected via 110kΩ (R5 + R6), whereas the inverting ('-') input is connected via 10kΩ (R7). This means that the bias current entering the '-' input will be greater than that entering the '+' input, deliberately creating a differential input offset voltage. Because of this, and the high gain of an op-amp connected without any negative feedback resistors, the output of IC1, VLED, will very nearly swing to the nega-

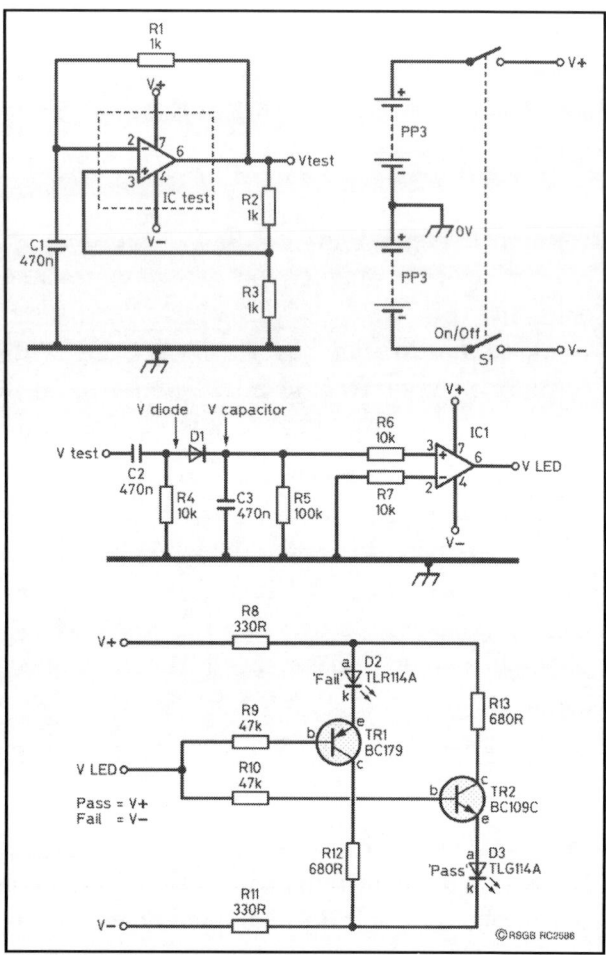

Fig 1. *Circuit diagram of the op-amp tester.*

tive supply voltage. This switches TR1 on and TR2 off, and so the 'Fail' LED D2 lights.

If, however, Vtest is a 1kHz square wave, C2 will not block the signal and D1 will conduct during the positive part of the waveform. This means C3 will charge up, as R5 is now acting as the discharge path for C3 and the time constant C3 × R5 is long compared with the period of the 1kHz signal. The resulting positive value of Vcapacitor causes the current flowing into the '+' input of IC1 to exceed the current flowing into the '-' input and VLED swings nearly to the positive supply voltage. This switches TR1 off and TR2 on, and so the 'Pass' LED D1 lights.

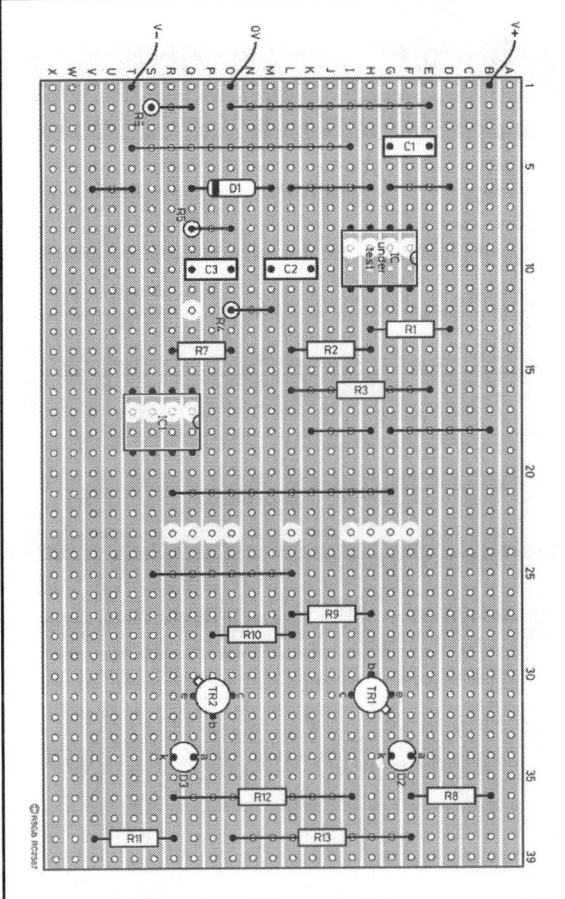

Fig 2. Stripboard layout and wiring diagram.

CONSTRUCTION

The stripboard layout for the op-amp tester is shown in **Fig 2**. An 8-pin DIL socket is of course needed for the op-amp under test, but IC1 can be inserted into a socket or soldered directly into place as preferred. It is always a good idea to install ICs in sockets though; as well as eliminating the possibility of damage due to soldering it makes removal for testing much simpler. Care needs to be taken to ensure that the op-amp, transistors, diode and LEDs are connected the right way round, and **Fig 3** shows how to determine the correct orientations.

COMPONENTS

All the semiconductors used in this project are general-purpose types, and the values of the components associated with them are chosen to limit voltages and currents to their working values. Otherwise, values of capacitance and resistance are chosen to give suitable time constants to ensure reliable operation.

IN USE

Any single op-amp package with standard DIL pin-out connections that operates with a power supply of ±9V can be tested, which covers most situations. Simply insert the op-amp to be tested into the test socket, again ensuring correct orientation, switch on, and note which LED lights.

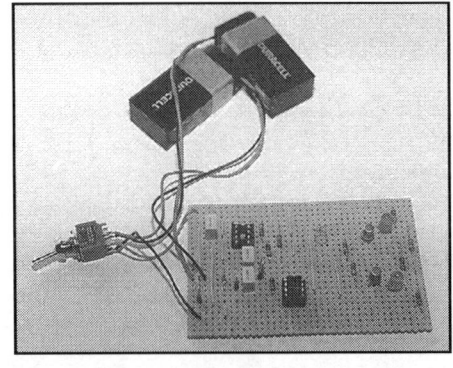

The completed op-amp tester.

USING A 12V SUPPLY

If the use of two 9V batteries to power the op-amp tester doesn't appeal to you, it is possible to adapt it to run from a 12V power supply. To do this, simply take two 100Ω resistors, make a potential divider (as shown in **Fig 4**), and use this instead of the top right-hand part of **Fig 1**.

PARTS LIST

Resistors - All resistors are metal film, 0.6W 1%

R1 – 3   1k

R4, 6, 7   10k

R5 100k

R8, 11 330R

R9, 10 47k

R12, 13   680Ω

Capacitors

C1 – 3 470n polyester film

Semiconductors

IC1   LM741CN

TR1   BC179

TR2   BC109C

D1 1N4148

D2 TLR114A red LED

D3 TLG114A green LED

Additional items

S1 DPDT

Stripboard

PP3 battery and clip – both × 2

A Useful Audio Level Indicator

Many amateurs, when wanting to try the many digital modes on offer, find that connections must be made from their transceivers to their computers. In the January 2004 RadCom, Ian White, G3SEK, explained how to make these connections, and gave advice on how to set the transmit audio levels so that overdriving and non-linearity were avoided. Here is an excellent piece of ancillary equipment to enable you to monitor the levels you set, thus ensuring complete repeatability when setting up.

In the January 2004 RadCom, G3SEK, in his 'In Practice' column, described the problems of setting transmit audio levels for PSK31 operation; a similar problem exists on receive for the NOAA and other weather satellites requiring software for the computer sound card. Several years ago, a useful level indicator was designed and built, which has been in use ever since, for several modes including PSK31 and WXsat.

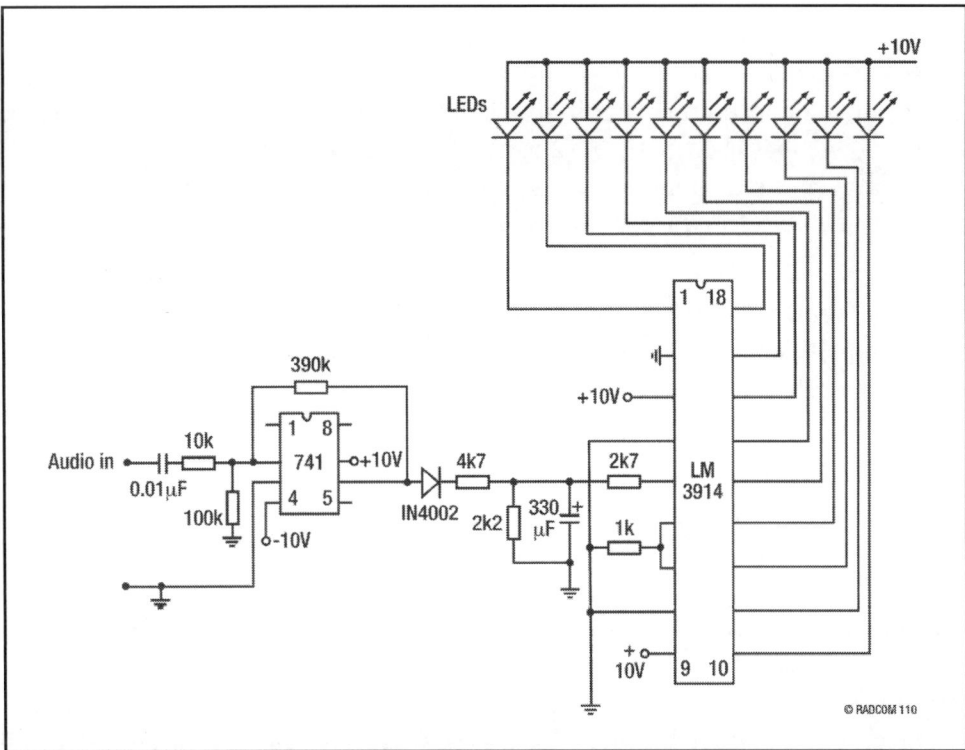

Fig 1. *The level indicator circuit.*

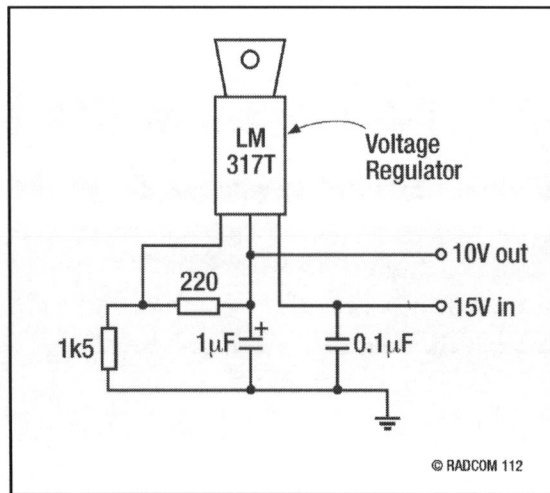

Fig 2. *Deriving the +10V regulated supply from a +15V unregulated source.*

WHAT DOES IT DO?

Connection of the indicator to the input terminals of the final piece of equipment in the chain enables the level for correct operation to be achieved easily. Receive calibration should be by the traditional hit-and-miss method, reviewing the results of each test until the optimum is found, the procedure being quite straightforward. Then you can make a note of the indicated level for future use. When using the device to monitor your audio transmit level, particularly for PSK31, use the technique described by G3SEK to set the level, and then use the indicator to register its magnitude, enabling you to set it precisely again when you have been using other modes.

THE CIRCUIT

The original indicator, the circuit of which appears in **Fig 1**, was built on a piece of stripboard about 5cm x 12cm (2in x 5.75in), and drew its power from the RIG RX2 receiver, but a simple power supply utilising a 12-0-12V transformer, two diodes and an electrolytic capacitor could be used.

Note that the 741 op-amp requires ±10V. This balanced supply is derived partly from the circuit of **Fig 2**, which provides +10V regulated from a +15V unregulated supply, and **Fig 3**, which produces -10V from the single +10V supply.

In the circuit, the audio signal is applied to the inverting input of the 741 op-amp, the output of which is rectified and, after integration, an LM3914 LED bargraph display driver is used to provide visual level indication by means of 10 discrete LEDs or a single bargraph module.

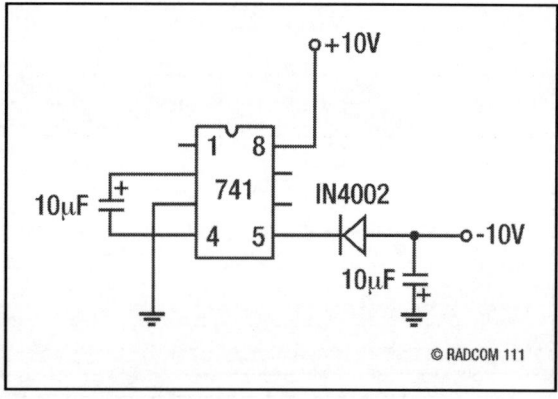

Fig 3. *-10V is generated from the +10V supply.*

AN ALTERNATIVE

A sensitive moving-coil meter could easily be used here, in place of the LM3914 and LEDs, although it has not been tried. The LED option is probably less expensive and is certainly very robust.

No claim is made that the design is an optimum one, but it worked as soon as it was switched on, and has proved to be completely reliable. The unit could probably be used as a training project for potential radio enthusiasts, as well as being a useful piece of gear around the shack.

Dual-Voltage Supply From One Battery

Many projects require a positive and a negative supply voltage, which can of course be provided by using two batteries or a dual power supply. A neater (and cheaper) method, however, is to generate both voltages from a single battery. One way of doing this is to use a simple voltage divider of two resistors but, unfortunately, this method is only really satisfactory if the circuit doesn't consume any significant current. This design allows current to be drawn without significantly affecting the supply voltage.

THE POTENTIAL DIVIDER

Fig 1(a) illustrates the principle of the potential divider. On the left are two 5kΩ resistors connected in series across a 10V power supply. There are no prizes for saying that the voltage at their junction is 5V. If we were to ground this middle point, we would have a dual power supply of ±5V. This only works, however, if neither side of the 10V supply is grounded.

To call this a dual power supply is a little misleading. Even though a multimeter connected across each resistor would read 5V, we cannot draw any power from the circuit without drastically affecting its output voltages.

Here is an example. In **Fig 1**(a), a load of 500Ω is placed across one half of the supply. The parallel combination of the 5000Ω and 500Ω resistors has an effective value of about 455Ω. This can now be considered to be in series with the other 5000Ω resistor across the 10V supply. This now draws, not 1mA as originally, but 1.8mA. (You can verify these figures for yourselves.) We now need to ask ourselves what is the voltage across the 500Ω load resistor? Well, the current is 1.8mA and the resistance is 500Ω, so the voltage is just over 0.9V. So you can see that this simple system cannot deliver any power into an external circuit. Because the original power supply is still delivering 10V, you can see that the negative supply (after connection of our load to the positive half) will now be 10V − 0.9V = 9.1V. Not only has the positive voltage virtually collapsed but the negative voltage has shot up to almost 10V! This system will not fulfil our purposes.

The situation can be improved somewhat if the potential divider is made from lower resistances, as in **Fig 1**(b). You may like to work out the corresponding voltages here, with and without the load. You will see that the voltage drop on connection of the load is smaller, but look at the current that flows down the divider chain − 100mA − even when there is no load!

This is a very inefficient system which wastes battery energy and of course shortens its life, and also needs resistors capable of dissipating the significant heat generated by the wasted power. If the circuit presents any significant current drain, for example to drive an LED or an audio amplifier, the supply voltage will drop significantly in one or both supply rails.

AN ELEGANT SOLUTION

An effective solution to this problem is to use the potential divider to generate a reference voltage at half the battery supply voltage, but to incorporate another circuit element that prevents the divider supplying any power. This extra element is an operational amplifier (op-amp) connected as a voltage follower, to provide the required load current at this voltage. In this way, the power for the load comes directly from the main power source through the op-amp, not from the potential divider.

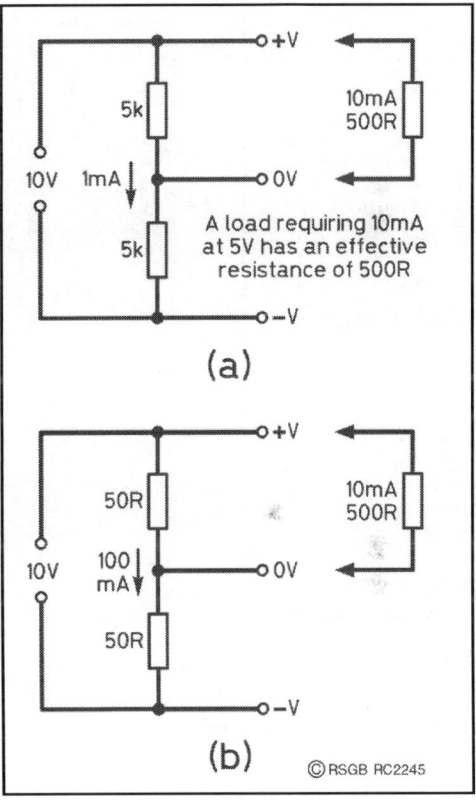

As before, we can use the output of the op-amp as our zero voltage level, and the two connections from the original power supply now become the positive and negative rails. This is shown in **Fig 2**. A battery is ideal as the power source because, as we noted above, neither side of it must be grounded if a true dual-voltage supply is required.

CIRCUIT DESCRIPTION

An op-amp connected this way is using 100% negative feedback, which means that all of the output voltage is fed back to the inverting input. This gives a voltage gain of unity, and means that the output voltage is always the same as the input voltage (hence the name 'voltage follower').

However, although it is called a 'voltage follower' it is actually a current-controlled device. A change in voltage at

Fig 1. The effect of a load on a voltage divider. (a) A voltage divider of 5kΩ resistors adequately provides V+ and V-of ±5V when there is no load, but if an effective load of 500Ω is connected across the upper resistor it effectively becomes 5kΩ in parallel with 500Ω, ie around 450Ω, and V+ drops to less than 0.5V. (b) If 10 times the required load current flows through the potential divider, ie the load presents 500Ω in parallel with 50Ω, the upper resistor is equivalent to about 45Ω and the loading effect is not so great. Even so, V+ will drop to around 4.5V.

the op-amp output caused by a change in load current also appears at its input due to the feedback. This causes the op-amp to change its output current in such a way as to counteract the voltage change on the output, and keep it fixed at the reference voltage. This change occurs effectively instantaneously, and means the circuit responds to all changes in load current demand while at the same time keeping the supply voltages fixed.

VALUES

The component values are not critical, and simple rules-of-thumb are sufficient for choosing values. The main thing to consider is the load current required, which must be within the specification of the op-amp. For example, a typical use of this circuit would be for powering an LM324 quad op-amp. This device will comfortably supply up to 20mA, which should be sufficient for most low-power applications.

The voltage divider resistors should carry around 10 times the input current for the op-amp, which is in the order of micro-amps, so any value

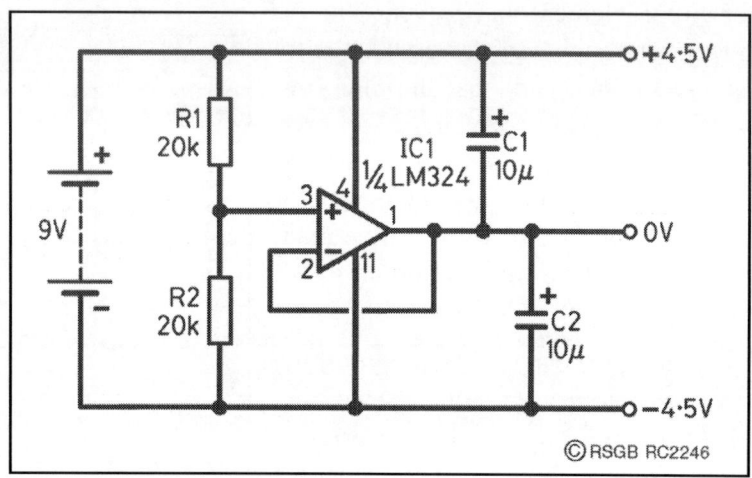

Fig 2. Circuit of the dual-voltage supply.

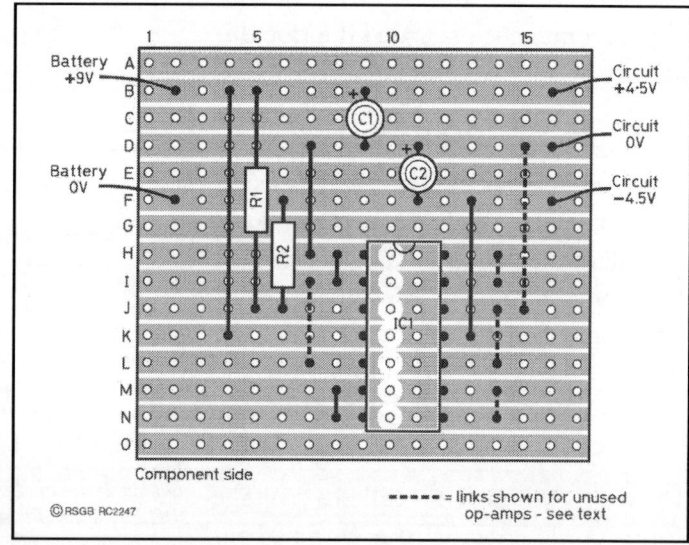

Fig 3. How to lay out the dual-voltage supply on stripboard.

between 10kΩ and 100kΩ will be fine. They do not need a greater power rating than that usually found in low-power circuits, so a 0.4W rating is more than sufficient.

The accuracy of the resistor values determines how closely the values of the plus and minus supply rails match, ie how close the reference voltage is to exactly half of the battery voltage. 1% tolerance is certainly adequate.

The capacitors are smoothing capacitors so, again, values are not critical, and could be increased to 100μF or even 1000μF without much increase in the physical size of the components. The voltage rating for the capacitors should naturally be higher than the output voltage.

STRIPBOARD LAYOUT

The circuit could be built as a 'stand-alone' project if required and used with different projects as needed. The component count is so low, however, that it's probably just as easy to build the power supply as part of a larger project.

The stripboard layout is shown in **Fig 3** for an LM324 quad op-amp, one quarter of which generates the plus, zero and minus supply rails which run along the full length of the stripboard for easy linking to other components.

The other three are available for use in the circuit of which it is part but, if these are not used, it is good practice to tie the output to the inverting input, and the non-inverting input to 0V. This

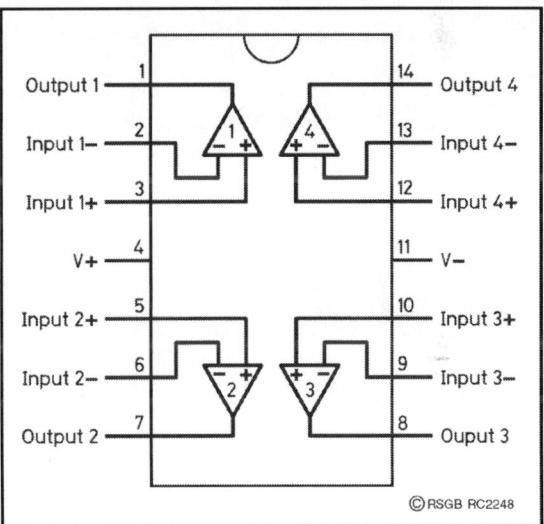

Fig 4. Orientation and internal layout of the LM324 quad op-amp integrated circuit.

reduces the possibility of any unforeseen behaviour of the op-amps such as latching or self-oscillation, which may increase current consumption or have other unpredictable and undesirable effects.

CONSTRUCTION

Construction is straightforward, the only things to watch out for being the correct orientation of the integrated circuit (check for the position of pin 1 – see **Fig 4**), and the correct polarity of the electrolytic capacitors (see the markings on their casings).

PARTS LIST

Resistors

R1, 2   20k, 0.4W, 1%

Capacitors

C1, 2   10µ (see text) electrolytic, voltage rating to exceed the supply voltage

Semiconductor

IC1   LM324 quad op-amp

Additional items

Stripboard

Battery clip

A Time-Out Unit For Digital Modes

The author has run digital modes for a couple of years and, aft... ...er of close calls, has designed a simple time-out circuit around the ... IC. The original setup used a transistor/relay arrangement to pr... ...tion between the computer and the transceiver PTT line and, for c... ..., the timer circuit has been added.

There is always, however, the possibility of the program or computer crashing with the PTT line remaining in the transmit state. This may not be too problematic in SSB mode if there is no modulation or if the operator is at the keyboard. But in Packet/APRS/UI-View modes, with FM modulation, and where the operators are, more often than not, absent for prolonged periods, the effect on the transmitter output stage can be catastrophic.

CIRCUIT DESCRIPTION

When the serial port goes positive (**Fig 1**), the transistor TR1 conducts, pulling the collector down to near 0V. When the serial port goes positive (**Fig 1**) the transistor TR1 conducts, pulling the collector down to near 0V, energising RLA. During the very short period before the RLA contacts actually make, the trigger terminal of the IC, pin 2, is also pulled down via the relay contacts RLA1, triggering the 555 timer period.

The output of the IC, pin 3, now goes to almost 12V, energising the relay, RLB, and closing the PTT circuit via the contacts, RLB1. The timer resistor Rt will now start charging the timing capacitor, Ct, and when the voltage at pin 6 of the IC

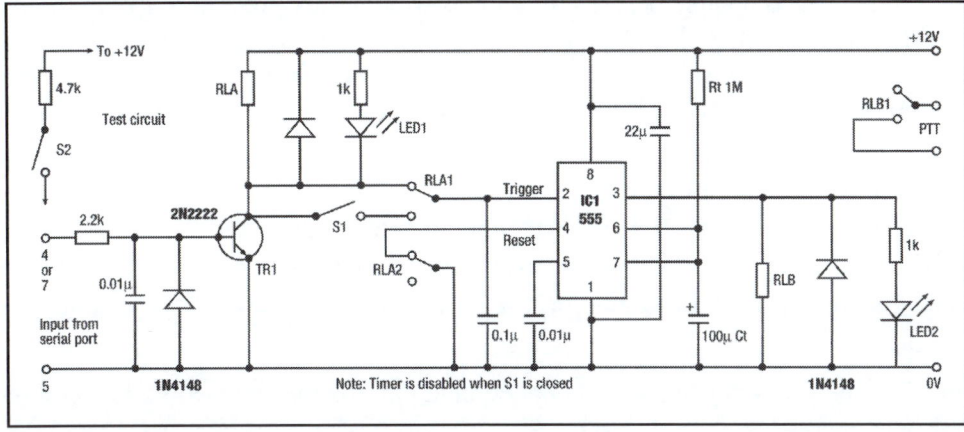

Fig 1. Circuit diagram of the time-out unit.

reaches an internal IC reference voltage, the output on pin 3 will fall to 0V releasing relay RLB and breaking the PTT circuit. When the serial port returns to 0V, relay RLA will be released bringing the reset terminal of the IC, pin 4, to 0V by way of contacts RLA2, which resets the timer. Similarly, if the serial port returns to 0V before the timing period has expired, RLA will be released, the timer will be reset, releasing RLB and breaking the PTT circuit.

Summarising, when the serial port goes positive, the transmitter is keyed until such time as the serial port returns to 0V or the timing period, set by Rt and Ct, has expired. The timer period is controlled by Rt and Ct, according to the expression Tp = 1·1 x Rt x Ct seconds. With the values shown in the circuit, the period should be about 110 seconds. In fact, the breadboard and two working models were both in the region of 120 seconds, no doubt due to tolerances of the capacitors. The timing period can be modified by the adjustment of Rt and Ct, but increasing Rt beyond 1MΩ is not recommended.

Beware of leaky capacitors that will have a significant effect on the timer period with such values of resistor. The LEDs are optional. LED1 indicates that the computer is signalling a 'transmit' condition and LED2 indicates that the PTT is activated. The switch, S1, when closed, turns the timer function off by holding the trigger terminal in a continuously-triggered state when the serial port is positive, ie signalling a 'transmit' state.

TESTING

The timeout period can be checked by connecting the input to the 12V supply via a 4.7kΩ resistor, such that it presents about 5V to the input terminal. When S2 is closed, both LEDs should glow and when the timing period expires, and LED2 should extinguish. Opening S2 will reset the timer and the process can then be repeated. If S2 is opened before the timing period has expired, both LEDs should extinguish.

DUAL-PORT OPERATION

Depending on the software, it is possible to run two separate programs on one soundcard by using both the left and right channels. The left channel being presented on the tips of the 'line in' and 'line out' stereo jacks and the right channel on the rings. You will probably find that microphone sockets are mono and, where dual-channel operation is required, it will be necessary to use a 'line' or 'auxiliary' input.

Pin 7 of the serial port is the left channel and pin 4 is the right. Pin 5 is common for both channels.

EARTHING

It is desirable to isolate the computer completely from the radio equipment. If this cannot be done, it may be preferable to bond down all equipment to the station earth.

PARTS LIST	Semiconductors
Resistors – all 0.25W	1 off 2N2222 npn transistor
1 off 2.2kΩ	3 off 1N4148 diode
2 off 1kΩ	1 off NE555N integrated circuit
1 off 1MΩ	2 off standard LED, any colour
1 off 4.7kΩ. For testing only	**Relays**
Capacitors – all 16V minimum	2 off miniature DPDT relay, 700Ω coil, 12V, Maplin UQ89W
2 off 0.01µF	Miscellaneous
1 off 0.1µF	2 off SPST switch. One is required but only for testing
1 off 22µF	
1 off 100µF	

Fed up of messing with a rats' nest of mixed colour wires, crocodile clips, test probes etc, I designed this simple 'breakout box' to make my life easier.

It is based on the fact that, in order to measure current, a circuit must be broken and a meter inserted. This is often mechanically awkward to arrange, so this box provides a convenient method of doing so. It also allows the use of unmetered or unprotected power supplies, and making connections with no more preparation than stripping the ends of some wires. The terminals can be to readers' own requirements.

The term 'breakout box' comes from computing and telephony systems, where inspection of a signal or data stream within a cable is needed. This unit enables the voltage and current being drawn from a power supply to be checked as adjustments are made, without disturbing the bench setup. A 50Ω BNC socket is included in order to observe on an oscilloscope factors such as ripple or noise on a DC supply. This impedance connector is used on test equipment, as well as for amateur radio. Please be aware that a mains-powered oscilloscope may well have its ground side (including the barrel of the BNC connector) connected to the mains earth, which may conflict with one or other of the DC supply lines in the test environment, except perhaps in battery-powered equipment that is isolated from other grounding systems.

Figure 1 gives the circuit. It was intended for my 0-30 volt power supply unit, but it is the mechanical design that is of note. It is semi-permanently connected to the output of my power supply (or, sometimes, batteries) and avoids wrong connections, slipping clips, ham-fisted and unreliable temporary solder joints, wires floating about with who-knows-what volts on them, risking damage to innocent semiconductors, shorts, other malfunctions etc. Of course, if you have 'real power' on the test leads, high voltages or current available, it is much better to make sure they are properly connected before you start! It would be advisable to include a fuse immediately after the input connector, and limit the breakout box to setups not exceeding say, 30 volts and a couple of amps.

Photo 1: General view of the breakout box with ammeter connected.

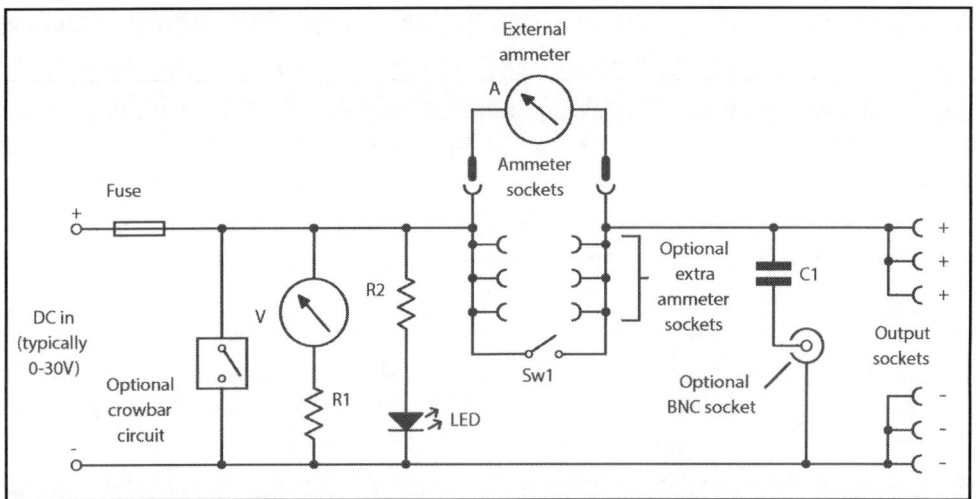

Fig 1: *Basic circuit of the breakout box. See text for component values and information on optional parts.*

The fuse should be selected appropriately for the capability of your power supply unit (PSU) or the circuit under test. If you have (say) a 30A PSU but are experimenting with circuits that won't need more than 1A or so, it makes a lot of sense to fit a 1A or 2A fuse. Self-resetting electronic fuses (eg 'Polyfuse') are useful, but make sure the voltage rating is high enough for your requirements. Remember that all the wiring inside your box – and the connectors – should be rated to carry at least as much current as the fuse rating, and preferably two or three times as much.

The unit provides instant, switched, ability to measure current, using the multimeter, whilst a voltmeter is wired in circuit permanently because it is cheap and easy to do. If your voltmeter is appropriate for the intended range, R1 should be replaced by a short circuit, but if the voltmeter range isn't quite right for your intended application (or you're using a microammeter in place of a voltmeter), you will need to calculate a value for R1 and fit it.

A current meter could be added too, but as one of this unit's specific purposes is to allow the insertion or withdrawal of a multimeter, on its current ranges, this is perhaps gilding the lily. A light emitting diode (LED) provides a power status indication. Current limiting resistor R2 should be calculated according to the intended voltage use – for 12V, 2k2 is about right; 4k7 will be fine for anything up to 30V or so, merely causing the LED to be dimmer at lower voltages.

SW1 has to carry the whole output current and must be rated appropriately. Miniature toggle switches of the type I used are readily available with ratings of up to 5A or so, but do check the rating before use because some may only be capable of 100mA or less. When SW1 is on, the external ammeter is bypassed; with the switch off, the ammeter is inserted into the circuit to read the current. Note that the ammeter comes after the voltmeter and LED.

Colour-coded, sprung loudspeaker connectors are used for Power In, Power Out and Test Meter points, with a variety of associated sockets for 2, 3 or 4mm test probes as required, depending on what equipment you have; both my meters use 2mm probes. Where stated by suppliers, the current carrying capacity of the speaker terminals appears to be a universal 3 amps, wherever obtained, and they will accept a meter test probe, or wire up to 2mm diameter (16 SWG). You may of course choose to incorporate other quick-connect terminals or binding posts, for example the type that takes a 4mm plug and screws down to accept a cable of up to 3mm or so. Provision for crocodile clips, scope probes, and sundry attachments, or perhaps spade or Lucar terminals may also be fitted as required. It may be advisable to make a small screen or shroud of insulating material between or around any exposed metal terminals. Plain perforated synthetic resin-bonded paper (SRBP) or Plastikard (a rigid styrene sheet, available from model shops) would be a useful material, fixed with small brackets or superglue. Or, one may leave such terminals off: this unit can be adapted to the reader's personal requirements and nothing should be regarded as a 'must fit' facility.

If you decide to fit a BNC socket to enable power line monitoring (for noise, etc) then you will need to decide whether to make it DC or AC coupled. DC coupling runs the risk of excessive current flow into an instrument or cable under fault conditions, so AC coupling is probably best. C1 provides this AC coupling – the value isn't critical, and anything from about 1nF to 1μF would probably be OK. 0.1μF is a good bet, but use whatever you've got in your junk box. Most capacitors have a sufficiently high voltage rating but be aware that some disc ceramics can be rated at 25V or less.

A 'crowbar' circuit is shown on the circuit diagram as an optional extra, discussed later.

CONSTRUCTIONAL DETAILS

Construction should be such as to suit readers' personal workshops. The author would be pleased to hear of additional ideas. A transparent polypropylene (Resin Identification Code PP5) 'hobby box' from a hardware shop formed the prototype's case, sprayed black inside with suitable automotive bumper aerosol paint. Unfortunately, my pressure-printed Dymo tape does not stick to this material, hence the lack of labels in the photographs. This plastic tends to crack or snatch on cutting and drilling, so some care is needed in machining this material. Enclosures made from other materials (eg ABS) are generally easier to work. A Forstner woodworking bit or a hole saw is useful when drilling the large hole for the meter.

Photo 2: Close-up of the prototype's connectors.

Brass nuts and bolts were provided on the prototype (not seen in the photos) for the convenient attachment of crocodile clips. The bolt heads should be uppermost, both to give more gripping surface and because, if installed 'thread uppermost', over time the threads may be damaged, leading to difficulties in removal or replacement. They are held in place by nuts inside and out, with a solder tag and shakeproof washers. As noted earlier, it is recommended that if you are going to have exposed terminals you should have some sort of shroud or cover to reduce the possibility of accidental short circuits – this is especially important if your PSU is capable of delivering a significant amount of current.

I haven't included any photos of the inside of my prototype because your build will probably be different from mine. As long as you use suitably rated components and thick enough wire you should be fine. There are no particular layout requirements. The photos show how I laid out the terminals on my prototype: you can, of course, alter this any way you wish to suit your own preferences.

POSSIBLE ENHANCEMENTS

In its current state, the breakout box can hardly be called 'electronic', apart from the LED, but there is scope for an 'electronic crowbar' or even a simple Zener diode clamp circuit to protect the circuit under test from overvoltage, as suggested on the circuit diagram. I won't present details here: crowbar circuits are easily found on the internet and usually consist of something like a Zener diode and a thyristor, with a few other simple components thrown in for good measure. Just make sure that the thyristor has a current rating greater than that of the fuse – a safety factor of two or three is useful.

More power conditioning such as a suitably-rated adjustable linear voltage regulator or pulse width modulated motor control circuit etc may also be added, each with its own set of terminals and separately switched, enhancing the usefulness of the box. Such pre-assembled 'building blocks' can often be obtained online at low cost, eg from eBay.

Geoff Theasby, G8BMI
geofftheasby@gmail.com

Arduino-based SWR analyser

In this project, a simple hardware SWR bridge is used as a measurement front end for the open-source Arduino development platform.

The user can develop their own solution based on the measurement software developed for the project, which is in the public domain and available on Github. It is not a complete, finished and ready-to-roll project: my hope is that this will stimulate fellow amateurs to build their own boards and experiment with the basic software to enhance the features that the analyser can provide.

In its basic form the analyser gives an accurate measurement of the SWR (in a 50Ω system) and impedance of an aerial at any frequency to >70MHz. Coaxial cable length can be measured and the resonant frequency of an aerial can be tracked. The analyser has the capability to be controlled from a program on a PC to allow data logging and/or graphical analysis of results.

SWR is a measure of the mismatch between the load and the design impedance of the system and gives an indication of the ratio of the impedances. This is a scalar measurement and does not give any information about the capacitive or inductive reactance except for its impedance at the test frequency. Judgement of whether a reactive load is capacitive or inductive requires interpretation of the measurement.

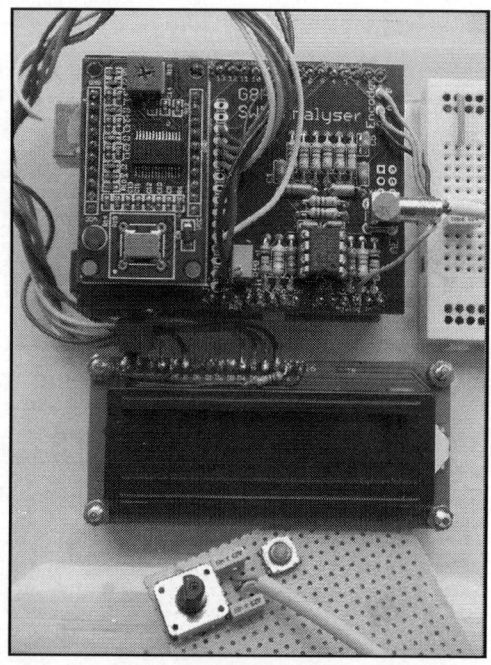

Photo 1*: The prototype board with piggybacked DDS module (left side of dark blue PCB).*

Project History

A desire for an inexpensive SWR analyser and an interest in direct digital synthesis (DDS) modules led to the start of this project. The AD9851 DDS is capable of synthesising a sinewave RF signal up to 50MHz or more with superb accuracy and stability. The cheaper AD9850 is good to 30MHz, but a small software change is needed to accommo-date a slight difference in control word for thelatter device. Both are available as assembled modules on eBay for only a few pounds.

The Arduino open-source platform uses a simplified C language environment for program development [1]. The ready availability of low cost digital hardware and a free software development platform to complement it allowed rapid evaluation of a proof-of-concept design.

The Arduino Uno board is based on an ATmega328, which has 20 inputs/outputs. Up to 5 inputs can be configured as analogue, feeding an internal analogue to digital converter (ADC) and two can become external interrupt inputs, which is perfect for this sort of project. The Uno motherboard has socket headers with all the sig-

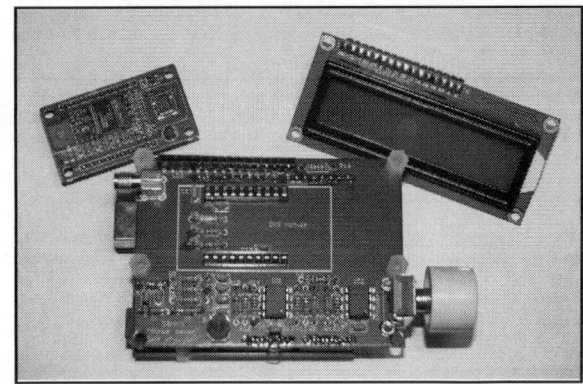

Photo 2: Second prototype. Main PCB (centre), DDS (top left) and display.

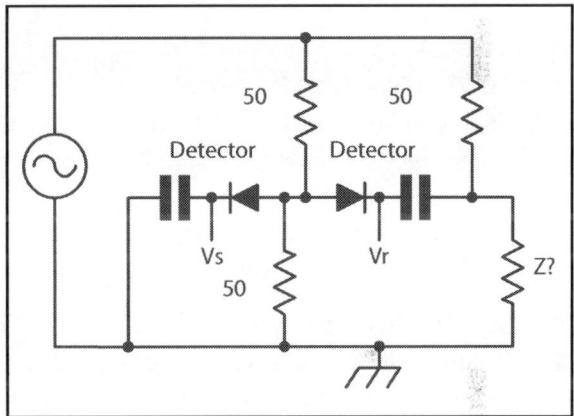

Fig 1: RF Wheatstone Bridge.

nals brought out to allow a development board to be piggy-backed. This gives a neat solution with plenty of resources to control the DDS, an LCD display and a rotary shaft encoder with push switch to select menu choices.

What's needed to make an SWR analyser?

For the majority of day-to-day practical uses, the amateur only needs to measure the SWR of the aerial at the chosen frequency, or to determine the frequency of lowest VSWR when tuning an aerial. In the past, this information was provided by a VFO, bridge, diode detector and moving coil meter: now, for a similar cost we can have a tool that can be programmed accurately to give greater flexibility and functionality (if the user so desires).

In the Wheatstone Bridge (see Figure 1), Vr gives an indication of mismatch

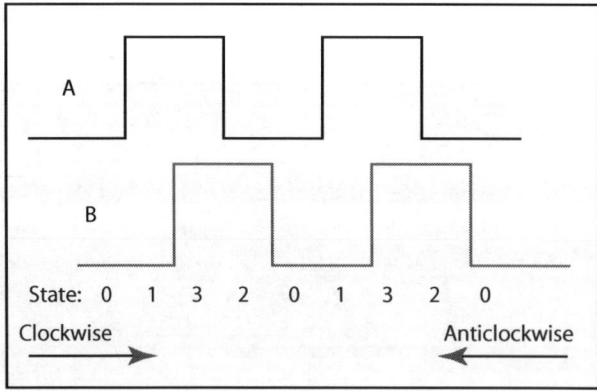

State: 0 1 3 2 0 1 3 2 0

Clockwise ⟶   Anticlockwise ⟵

Fig 2: Level transitions of quadrature encoder.

and would traditionally have driven a moving coil meter. The ready availability of low-cost microcontrollers such as the Arduino allows us to convert Vs and Vr to digital values and process them. Most commercial aerial analysers for the amateur market use some form of the Wheatstone Bridge.

A calculation involving Vs and Vr will reveal the SWR:
$$VSWR = (|Vs|+|Vr|)/(|Vs|-|Vr|)$$
where $|V|$ is the absolute value of the voltage, ie the magnitude.
If the bridge is in balance ($Z? = 50\Omega$),
$Vr = 0$ and $VSWR = 1$
If the load is open circuit, $|Vr| = |Vs|$ and $VSWR = (2xVs)/0 = \infty$
If the load is short circuit, $|Vr| = |Vs|$ and $VSWR = (2xVs)/0 = \infty$

A diode RF envelope detector (as shown in **Figure 1**) allows the magnitude of the RF signal to be measured accurately down to millivolt levels (when lightly loaded, eg buffered by a high impedance amplifier circuit). At low signal levels the diode exhibits a logarithmic response, but this can be compensated for in software. A comparative measurement of the I/V curves of 1N914 silicon small signal, 1N34 gold bonded germanium and BAT43 Schottky diodes revealed that the modern Schottky diode performs very well, with a forward voltage drop of only 23mV at <1µA. And, unlike the older 1N34, BAT43 diodes are readily available!

The detector that measures Vr measures the magnitude of the voltage difference between the two arms of the bridge, and truly floats at AC whilst producing a ground referenced DC signal through the return path consisting of the bridge resistors.

Each low-level detector output is buffered by a high input impedance operational amplifier circuit with sufficient gain (around 20) to give a useful output in the 5V range of the ADC on the AVR processor chip. Following digitisation, the non-linearity of the diode detector is compensated for in the software to improve the accuracy of the results.

The same VSWR will result from both high and low load impedances: SWR 2:1 for both 100Ω and 25Ω, SWR 3:1 for 150Ω and 17Ω, etc. The addition of a third detector to measure the voltage across Z gives us the opportunity to determine whether the load impedance is greater or less than 50Ω. If Vz is low, the impedance must be low; a high Vz is the result of a high impedance.

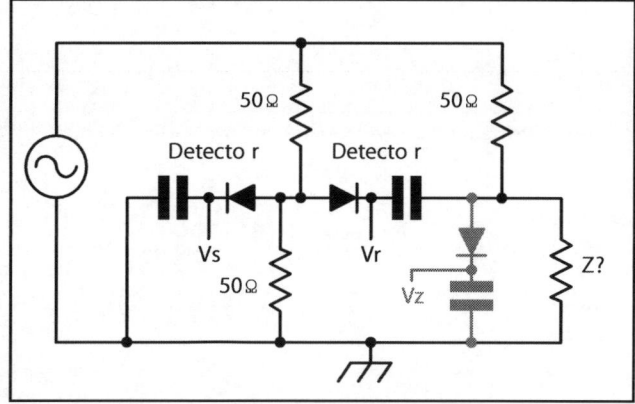

Fig 3: Addition of detector for Vz (in tint).

From breadboard to prototype

Work on a basic two detector bridge design using a breadboard proved that the concept was sound and allowed me to write the base software to control the DDS, accept input from the shaft encoder and drive a two line LCD display for menu selection and to display results. I then designed a prototype PCB to allow further development in a more controlled environment.

OSH Park [2] in the USA offer an excellent service for cost-conscious experimenters and provided me with 3 PCBs for $5 per square inch, delivered in roughly 2-3 weeks. The painless process involves submitting the design file from a PCB CAD package to their website and paying by PayPal. The designer is kept fully informed on the progress of the PCB, and following the announcement that it has shipped the excitement grows! (Other prototype PCB manufacturers offer comparable terms – Ed).

The board is designed to plug directly into an Arduino Uno. The Arduino requires a stable 5V power supply, which can either come from a PC connected to the Arduino's USB port or from an external low voltage DC supply (eg a 12V wall wart or batteries) through the on-board regulator. The Arduino has a USB serial link to an attached PC that is used for downloading the program to the Arduino as well as being a communications link to allow data logging and debugging when the analyser is in action.

In **Photo 1** the DDS module can be seen piggybacked on the left hand side of the prototype PCB, which itself is plugged into the (hidden) Arduino Uno development board.

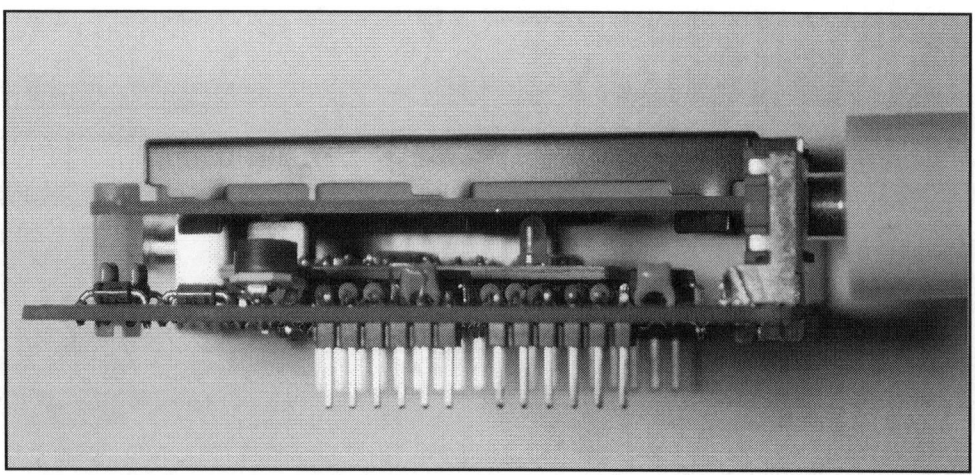

Photo 3: *The stacked PCB, DDS and display.*

The DDS is controlled through a Serial Peripheral Interface (SPI), which uses a clock, data in and data out signals to transfer the control word between the processor and the slave DDS device at very high speed. The control word of the DDS is programmed with a 32 bit number to define the required frequency, which can be set to an accuracy of 0.04Hz(!)

The quadrature shaft encoder uses two signals (A and B) that alternately go high/low/high as the shaft is rotated, but with a lag between them. **Figure 2** shows how this works.

The order in which the A/B quadrature signals are received is interpreted by the processor to give the direction of rotation and keep count of the relative position of the shaft. Using interrupts to detect the level changes allows the processor to continue with foreground tasks whilst the encoder is being rotated: each increment of the encoder creates an interrupt that calls the service subroutine, only taking a few microseconds to update the shaft position. My shaft encoder interrupt service routines were developed from code originally written by Beric Dunn, K6BEZ.

As the shaft is rotated, the processor counts with the transitions and increments or decrements the position count, which serves to index through menu options or control the DDS frequency. A separate switch contact made by pushing in the shaft of the encoder signals to select the current option. The menu structure is discussed later.

The success of the first board, plus the limitations of having the display and shaft encoder connected by cables, prompted the design of a second PCB (**Photo 2**, **Photo 3** and **Photo 4**). Originally, SMA connectors were chosen because of their small foot-

Photo 4: *Analyser with 50Ω termination.*

print, but were replaced with BNC in the latest revision.

As mentioned earlier, the simple two-detector measurement of SWR has the limitation that it cannot distinguish between a low or high impedance load, both of which exhibit the same mismatch (eg 25Ω and 100Ω both give SWR of 2:1). In order to judge the magnitude of the load impedance, another detector is needed to measure the voltage Vz developed

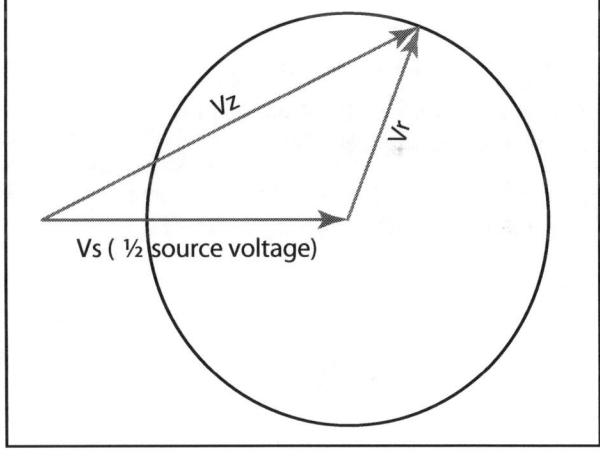

Fig 4: *Vector triangle for a reactive load.*

across the load, as seen in **Figure 3**, and this was added on the new board. A low value for Vz implies a low impedance and a high Vz a high impedance.

Note that all the voltage measurements are magnitudes and convey no information about phase shift in a reactive load, but the measurement of Vz gives us a clue to the magnitude of the reactance by solving the triangle of voltage magnitudes formed from Vs, Vr and Vz (see **Figure 4**). The core code is shown in **Figure 5**.

The story so far

The hardware is at a stage now whereby measurements of Vs, Vr and Vz can be made and digitised, there is a control interface consisting of shaft encoder, switch and LCD display and a DDS delivers RF to the bridge at up to about 50MHz or more, with precise control.

The rest of the project is software, which is open-source. At present, the menu system (**Figure 6**) allows the following user control:

```
// VSWR
void VSWR() {
  double Vs=0;
  double Vr=0;
  double Vz=0;
  double VSWR = 0;    // VSWR = (1+Γ)/(1-Γ) = (Vf+Vr)/(Vf-Vr)

  // Read the forward and reverse voltages
  Vr = analogRead(REV_analogue_pin);
  Vs = analogRead (FWD_analogue_pin);
  Vz = analogRead (ZL_analogue_pin);
  if (Vr >= Vs) {
          Vr = Vs - 1;
          }   // To avoid a divide by zero or negative VSWR

  // Calculate VSWR
  SWR = (Vs+Vr)/(Vs-Vr);

  // Linearise VSWR
  SWR = 1+9.55*log10(SWR);

  // Calculate load impedance
  if (Vz <  Vs) {         // ZL is low
      ZL = 50/VSWR;
  }  else {               // ZL is high
      ZL = 50* VSWR;
  }
}
```

Fig 5: Core code.

- Select any of the HF bands as a base to manually control the frequency of measurement and display VSWR and load impedance

- Measure the length of an open circuit coaxial cable

- Find and track the frequency of best match (useful for tuning an aerial).

Next steps

There are applications for the PC written in Python script (eg by PA2OHH) that provide a graphical interface to display measurement results from the likes of this board, with control and data being exchanged over a serial USB connection. The next work to be done will involve additional software for the SWR analyser to accept connection to and control by the PC to provide a remote measurement head that can operate via a convenient USB connection.

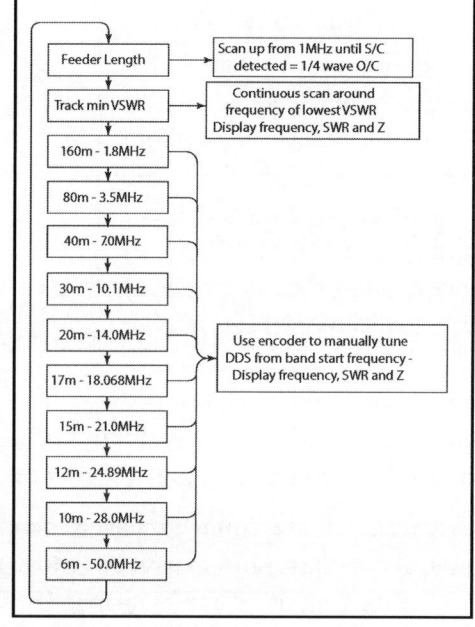

Fig 6: Menu structure

As mentioned earlier, the impedance of a capacitive load decreases with frequency, so Vz for a capacitive load would also decrease (and the opposite is true for an inductive load). Therefore a trick to determine the sign of the reactive component of the load is to have the software solve the triangle to calculate the magnitude and phase of Z? and then 'jiggle' the measurement frequency and determine the direction of change (and hence sign) of the complex part of Vz.

Including and beyond this, development of the software is open to the user. I will be very interested to hear what you do.

Michael Booth, G8HKS
mike.booth@shirenet.co.uk

Websearch

[1] www.arduino.cc

[2] www.oshpark.com

Further reading

By conincidence, this month's Homebrew takes a separate look at the RF Wheatstone bridge – see p36

Measuring RF Impedance Using the Three Meter Measuring Bridge, Peter Dodd, G3LDO, ARRL Antenna Compendium, 1995

Three Meter Method, Peter Dodd, G3LDO, RadCom September 2005

Amateur Measurement of R+jX, Doyle Stradlund, W8CGD, QST June 1965

I am a bit deaf and at special events I can have trouble hearing incoming audio. This also happens when others are speaking nearby, for example at the club shack. So I tend to wear headphones – which are great for me but shut everyone else out from the receive audio. Not quite what you want when running a demonstration station.

The answer we found is the very simple audio splitter box described here. It takes high-level audio from the loudspeaker output of the radio and splits it into four (or more) mono outputs, each with its own volume control for convenience. In order to maximise flexibility, we put both 3.5mm and 1/4" (6.35mm) sockets on each output.

The circuit diagram is shown in **Figure 1** and **Photo 1** shows the internal construction. The device was built into an interestingly-shaped case from the junkbox that already had one 1/4"

Photo 1: Inside the audio splitter showing two banks of variable resistors and socket sets (see text).

socket at one end. We added another on the other end, so that audio could be looped straight through. It was convenient to add eight outputs in this enclosure. The larger, metal-cased variable resistors at the bottom of Photo 1 are 100Ω and each is connected in parallel across the audio input. The wiper of each variable resistor feeds paralleled 3.5mm and 1/4" (6.35mm) mono sockets so that either size of plug can be used on each output channel.

As the variable resistors are driven directly from the loudspeaker output of the radio they act as four 100Ω resistors in parallel when no headphones

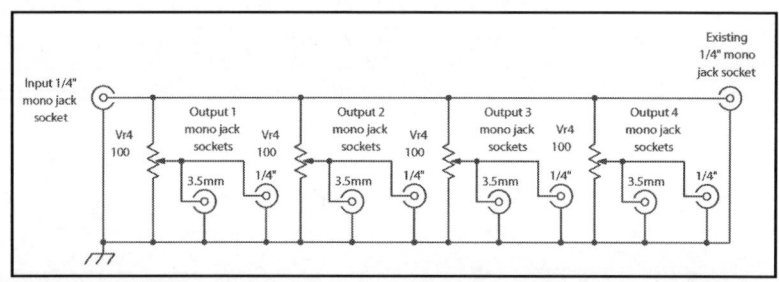

Fig 1: Circuit diagram of the audio splitter. Four output channels are shown but this can be extended as required. For (additional) high impedance 'line level' outputs described in the text, use variable resistors of anywhere between 1kΩ and 100kΩ or so – the value isn't critical.

etc are connected, meaning that the radio sees a load of 25Ω. When headphones or speakers are connected, the radio sees a lower impedance that depends on what's plugged in and the position of the volume controls. Most modern headphones are typically about 32Ω so if four sets are plugged in and all the volume controls are at maximum, the minimum load presented to the radio will be about 6Ω. This will be significantly less if lower impedance headphones, or loudspeakers, are used. Note that the variable resistors are only fairly low power devices so the audio splitter should not be used with audio sources capable of delivering more than a couple of watts or so.

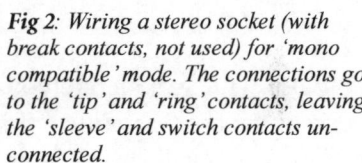

Photo 2: Outside view of the completed unit.

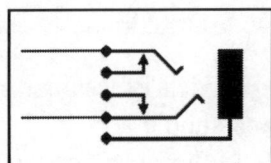

Fig 2: Wiring a stereo socket (with break contacts, not used) for 'mono compatible' mode. The connections go to the 'tip' and 'ring' contacts, leaving the 'sleeve' and switch contacts unconnected.

Our prototype (Photo 2) also contained four higher resistance, physically smaller variable resistors (at the top of Photo 1) that are coupled to 1/4" mono sockets. This lets the device connect 'line level' equipment to the audio system. We have used this unit very successfully with a bhi noise eliminating speaker with internal amplifier when running demonstration systems: the operator can hear the full range of signals and noises, which their experience will let them filter 'by ear', whilst the public benefits from the noise reduction provided by the bhi system.

The great thing about the audio splitter is that I can use headphones at whatever volume setting I want without significantly affecting the audio fed to others. I get to hear just the radio audio, without distractions, whilst others can engage with visitors to explain what's going on. Once you get it all set up it works very well.

As described, this system uses mono sockets throughout. If you prefer you can wire up stereo sockets in 'mono compatible' mode, connecting just to the tip and ring, as shown in Figure 2. This will mean that stereo or mono headphones can be plugged into the socket and will work just as well. But be aware that you shouldn't use this method to connect to a stereo device such as an amplifier or sound card using a stereo lead because the wiring will be wrong. A mono to stereo lead should be used in this instance, with the mono side plugged into the splitter.

This project was developed by members of the Bedford and District Amateur Radio Club, with particular help from Vince Maund, G8CZP.

Glenn Loake, G0GBI
glen.loake@ntlworld.com

The performance of any receiver or transceiver, no matter how expensive it is, is limited by the aerial that feeds it. Two of the most frequently asked questions are:

- Which is the best sort of aerial to use?

- Where is the best place to locate an amateur radio aerial?

To answer these questions, you must ask yourself what sort of operation you want to do. Are you interested in local, chatty contacts on the lower bands or VHF, or are you more disposed towards long-distance (DX) contacts, and on what band?

A house with a moderately-sized garden is assumed in the diagrams here, to illustrate the configurations of some simple aerials. You would not need all these aerials festooned around your house, because one or two would be sufficient for your needs. The problems incurred by properties with more restricted space will be/covered later.

VHF AERIALS

For VHF operation, the aerial should be mounted as high as possible, either on a mast or on a chimney. For all-round coverage on FM and the local repeaters, a vertical colinear is a good choice. For SSB and CW DX operation, a horizontal rotatable beam is needed. If satellite working is envisaged, you will need to contemplate mounting an elevator on top of your rotator, so that your beam can point in any direction, including vertically upwards! An advantage of satellite working is that the aerials do not necessarily have to be up in the air, provided you have a relatively uncluttered site. Your rotator and elevator can be at ground level, which is good!

If the VHF aerial is mounted on the chimney, use a double mounting bracket, particularly if you have a beam and rotator. Keep the TV, broadcast FM and amateur aerials as far apart as possible, and keeping the feeders separated is also a good plan.

THE DIPOLE AERIAL

One of the simplest types of aerial for single-band operation is the half-wave dipole. (The name 'dipole' simply means 'two poles' or 'two elements', and in this case the total length of the dipole is approximately half a wavelength at the operating frequency.) It is usually fed in the centre by coaxial cable as shown in **Fig 1**. The length of the dipole for the lowest frequency in each band is shown in **Table 1**. Normally, the length of the aerial will be 'trimmed' to be tuned to the centre frequency of the part of the band in which you will operate. This is done using the data in the right-hand column of **Table 1**. As an example, suppose you

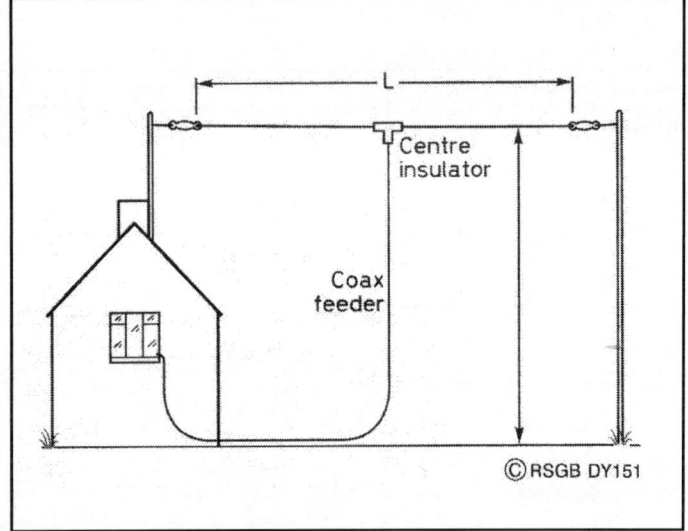

Fig 1. Layout for a dipole aerial.

wanted your aerial to be resonant at 3.7MHz. The table gives an overall dipole length of 42.86m for 3.5MHz. To resonate the aerial 200kHz higher, then this length must be shortened by 2 x 0.595 m = 1.190m. Your dipole would thus be 41.67m long. Remember to allow extra wire for fixing the dipole ends to the insulators.

Band (MHz)	Dipole length (m)	Trim each end (mm/10kHz)
1.8	83.33	2190
3.5	42.86	595
7	21.43	150
10	14.85	70
14	10.71	35
18	8.33	20
21	7.14	15
24	6.03	12
28	5.36	10
50	3.00	6

Table 1. Dipole lengths for lowest frequency of each band and the length to be trimmed from each to raise the resonant frequency by 100kHz.

On the lower-frequency bands, the lengths become rather large. In this case, you can 'bend' your dipole, as illustrated in **Fig 2**. The length of wire required to give an acceptable value of SWR (less than 2:1 on transmit) may need to be different from the calculated value, so be prepared to experiment!

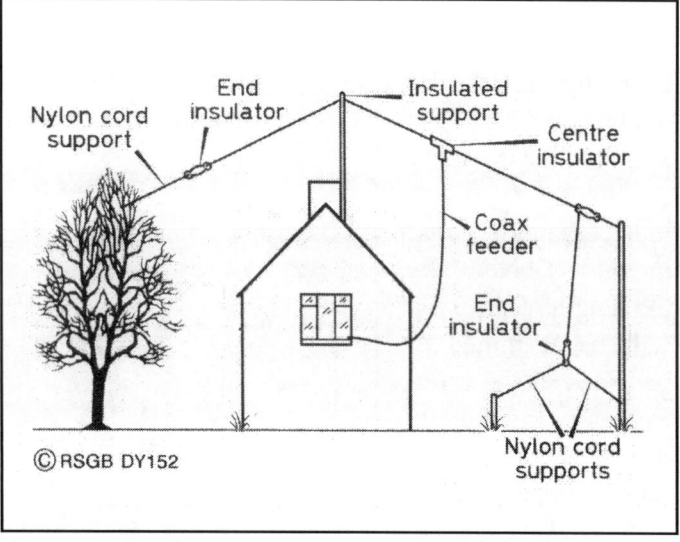

Fig 2: Possible layout for a dipole aerial in a confined space.

Dipoles are single-band aerials, although they will often work acceptably on the third harmonic of their design frequency – a 7MHz dipole often operates reasonably well on 21MHz. It is possible to operate several dipoles in parallel, as **Fig 3** shows. Interaction between the elements can occur if the spacing between them is less than about 10cm. A multi-band dipole, as shown in **Fig 3**, has the elements separated with plastic spacers, and drooping ends to produce maximum spacing between the elements' ends.

THE LONG-WIRE AERIAL

This aerial is simple, cheap, easy to erect, and suits most houses and gardens, as **Fig 4** shows. Using an aerial tuning unit (ATU), an end-fed long wire can function on several bands when used with a set of radials or a counterpoise. **Fig 5** illustrates

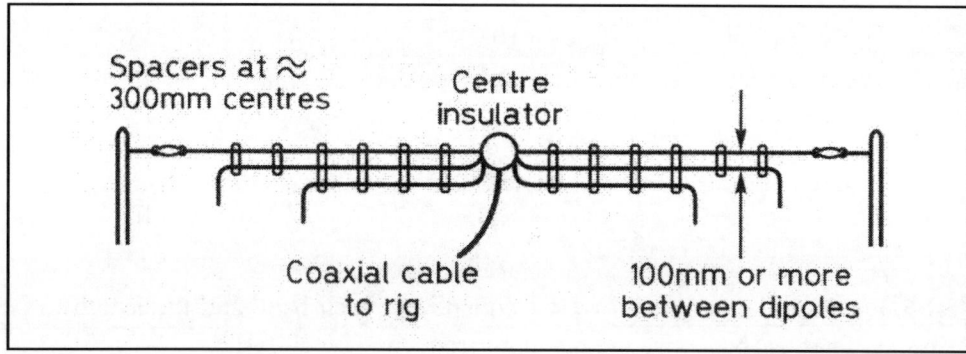

Fig 3: Multi-band dipole aerial.

the setup. The length of the aerial will determine the bands which will be covered.

A wire length of 10.5m will work on the 40, 30, 17, 15 and 12m bands.

A wire length of 15.5m will work (with an ATU) on the 80, 40, 20 and 12m bands and possibly (depending on your ATU) on the 17 and 15m bands.

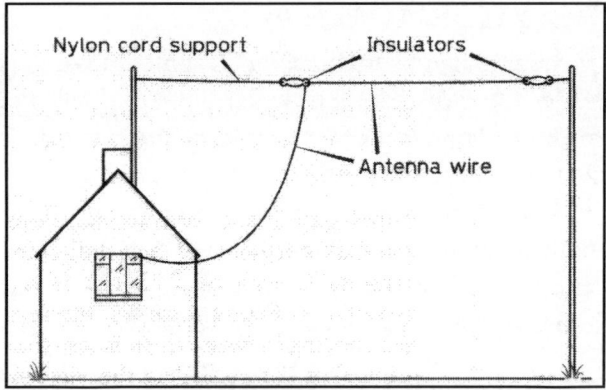

Fig 4: Long-wire or inverted-L aerial.

A wire length of 26.5m will operate on all bands, but may be difficult to load on 10m.

The wire lengths given here may need some adjustment because of the geometry of your particular house and garden. For receive-only purposes, the lengths are far less critical.

In general, you cannot get a good radio-frequency (RF) earth from a first-floor (or higher) shack. Unless a good RF earth exists within a small fraction of a wavelength from the transceiver, an artificial ground comprising a single $\lambda/4$ radial or counterpoise will be needed. You will need one counterpoise for each band you intend to use, and the wire can be concealed around the skirting-board of the shack, or under the carpet. Make sure that the free end of each counterpoise is well insulated; this point can carry a very high voltage when you transmit; anyone coming into contact with this can suffer very severe RF burns. Counterpoise lengths can be read from **Table 2**.

Band (MHz)	Element length (m)
1.8	39.66
3.5	20.40
7	-10.20
10	7.14
14	5.1
18	3.96
21	3.4
24	2.99
28	2.55

Table 2. lengths of elements for vertical antennas, radials for verticals and counterpoises for end-fed long-wire antennas.

THE VERTICAL AERIAL

The single-band vertical aerial is sometimes used by DX operators because it has a low angle of radiation, which favours long-distance propagation. However, it must be sited clear of obstructions and must have a good counterpoise or radial system. Illustrations of the vertical aerial are shown, and the lengths of the vertical and radial sections are given in **Table 2**. The centre of the coaxial feeder is connected to the vertical section, and the braid to the counterpoise or radial system, which is made up of four or more wires buried just below the surface and joined together near the base of the aerial.

CABLE ENTRY TO THE HOUSE

Bringing coaxial cable into the house by an open window must be regarded as a temporary measure. Wooden window frames can be drilled, one hole for each feeder. Make the holes slope downwards from inside to out to prevent rain entering, and treat these with wood preservative. Leads from long-wire and inverted-L aerials should be kept separate from other cables.

Alternatively, a plastic pipe large enough to take all your feeders could be fitted into the brickwork (again, sloping downwards towards the outside). You may want to let a friendly builder do this for you.

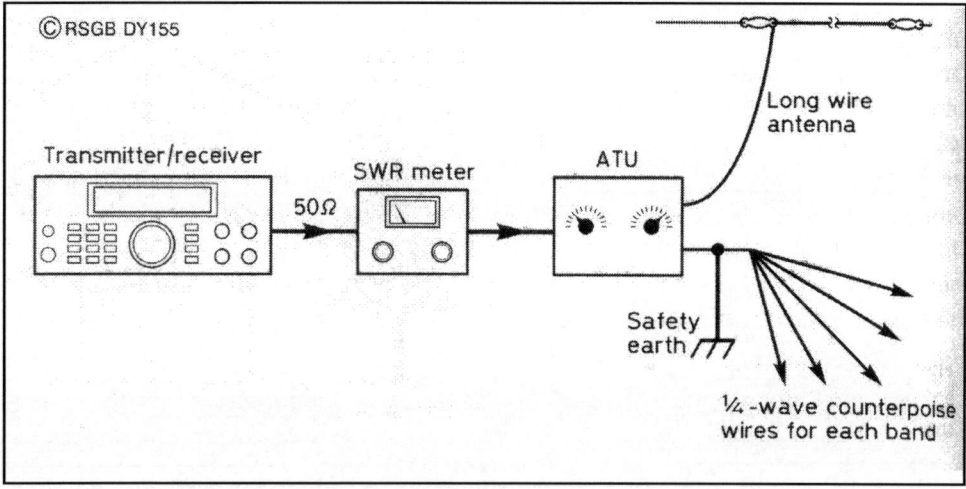

Fig 5: *How to connect the radio to a long-wire aerial.*

Many people have transceivers with digital signal processing (DSP) inside them. Here is a simpler approach, based on an analogue integrated circuit.

The prototype used some surface-mount devices (SMDs) but this is not necessary to enjoy the benefits of this little circuit.

DSP

The test reports on digital signal processors promise miracles; many people feel that the noise is reduced in the noise reduction mode but so is the readability!

The Analog Devices SSM2000 IC [1], works on the HUSH principle, developed and patented by Rocktron Corporation. Although designed for the hi-fi market, it has found an application in amateur radio which is described here.

HOW IT WORKS

The HUSH system can distinguish between signal and noise because the volume and frequency spectrum of speech or music continually changes; by contrast, the

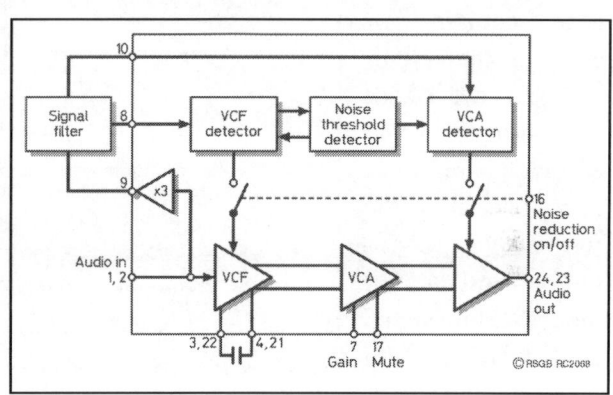

Fig 1: *Block diagram of one channel of the Analog Devices SSM2000 IC.*

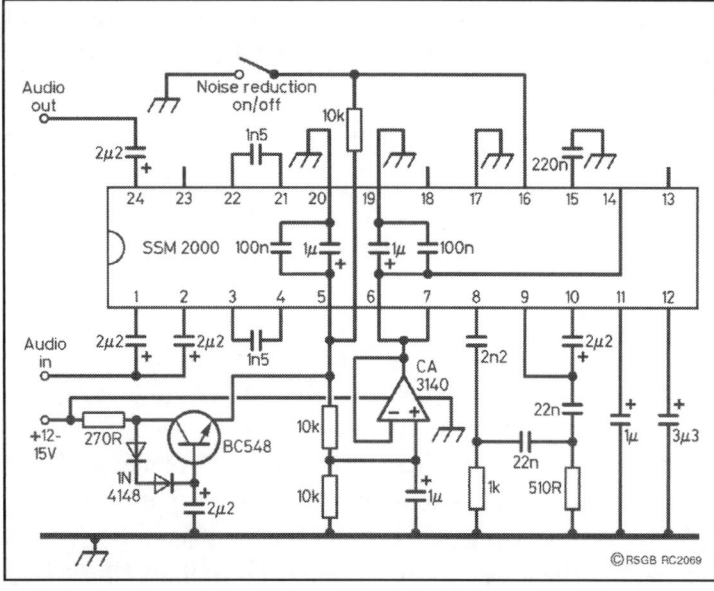

Fig 2: *The noise reduction circuit, for use in a communications receiver.*

noise amplitude and frequency spectra remain relatively constant.

Following the block diagram in **Fig 1**, the audio signals (only one channel of the two required for stereo sound applications is described) are processed to extract information concerning the frequency distribution and amplitude of both signal and noise, passing through low-pass voltage-controlled

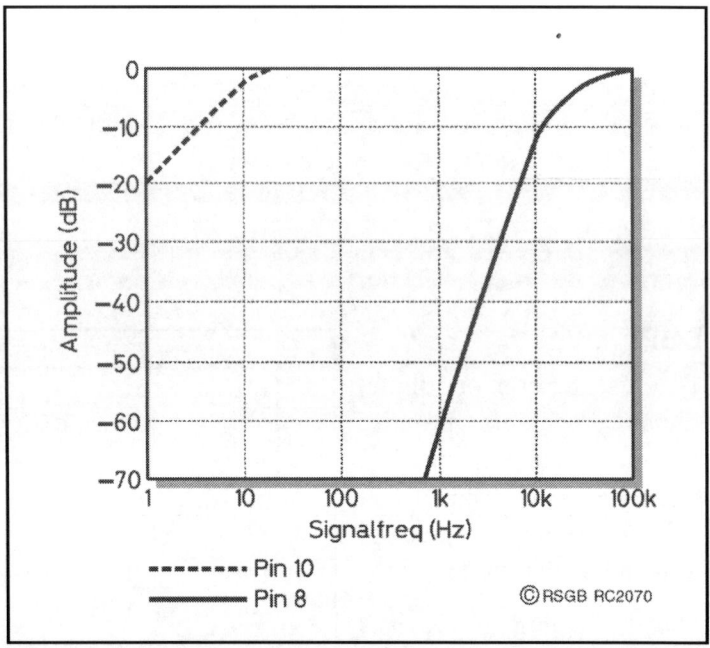

Fig 3: Low-frequency roll-off to pin 8 through the 3-pole filter in Fig 3. Also shown is the signal on pin 10, which does not go through the filter.

filters (VCFs) and then through voltage-controlled amplifiers (VCAs). In contrast to digital processes, VCFs and VCAs have low distortion and add negligible noise of their own. The cut-off of the VCF and the gain of the VCA can be set as required by the application. With control signals derived by a proprietary algorithm and applied to both VCF and VCA, a noise reduction of up to 25dB can be achieved.

THE APPLICATION

Looking at the diagram in **Fig 2**, the signal is applied to pins 1 and 2. It comes out of pin 9, amplified by a factor of three, from where it is fed unaltered to the VCA detector (pin 10) and through a three-pole high-pass filter to the VCF detector (pin 8). **Fig 3** shows the two frequency responses, optimised here for hi-fi music; for narrow-band voice signals it would be appropriate to move the frequency response of the three-pole filter downward by making the capacitors about five times bigger (22nF to 100nF, 22nF to 100nF and 2.2nF to 10nF respectively). The time constant for the decay of the control voltage to the VCFs is set by the 1µF capacitor on pin 11. The noise threshold is 'held' on the 220nF capacitor on pin 15. The control voltage to the VCA decays according to the value of the capacitor on pin 12. The lower cut-off frequencies of the VCFs are determined by the 1.5nF capacitors between pins 3 and 4, 21 and 22.

The supply voltage on pin 5 is decoupled by the BC548 emitter follower. The op-amp holds pins 6, 7 and 14 on half the supply voltage. The switch on pin 16 allows the noise suppression to be disabled.

RESULTS

The unit was built on a double-sided 52 by 33mm PCB. Except for the DIL ICs, SMD components were used throughout. Through-connections are by way of the terminal pins and the pins of the DIL sockets, which are soldered on both sides. The two 1.5nF capacitors were soldered directly to the IC.

There are no adjustments. As an audio input of 0.3V is recommended, the noise suppressor was installed between the product detector and the volume control of the receiver, an old Drake R4-C. The product detector output is fed to both the L and R inputs of the SSM2000. The sum of these inputs may be applied to pin 9 of the VCA and VCF controllers. The output to the volume control is taken from one channel only.

The manufacturer claims a noise reduction of up to 25dB in hi-fi systems. This cannot be obtained with speech, but the performance is very pleasing as it is achieved without audible distortion of the wanted signal.

REFERENCES

[1] A 16-page data sheet for the SSM2000 can be downloaded from www.analog. com.

Mention microwaves and many beginners will say "It's too complicated, too costly and too difficult for me". Well, present-day, state-of-the-art, narrow-band equipment designs almost certainly are too difficult and possibly too costly for beginners. Many newcomers do, indeed, appear to be put off by this high technology.

So where do beginners start? This article may help to revive interest in the simple, easy, and inexpensive wideband FM (WBFM) designs that most of today's advanced operators used before the latest state-of-the-art designs became available for experienced home constructors.

The ideas are not new, so don't expect a 'blow-by-blow' description. Full details are available in the references quoted*. The intention is to encourage you to get hold of this information, have a go, produce working equipment, get the feel for microwave operating and then, perhaps, to have a go at some of the more advanced microwave designs as your skills and knowledge improve with practice. Believe me, you don't have to be a skilled constructor or operator to take the first steps.

WHAT'S NEEDED?

You will need the following major 'ingredients' to construct a 10GHz WBFM transceiver:
- An aerial (see, for example, the horn aerial in reference [2], pp34 – 41)
- A mast (for fixed station working) or tripod (for portable working see reference [2], pp107 – 108)
- An in-line Gunn oscillator/mixer Doppler intruder alarm module (see reference [1], pp98 – 102)
- A Gunn oscillator power supply/modulator module (see reference [1], pp94 – 97)
- A low-noise, wide-band IF preamplifier module (see reference [2], pp122 – 124 or pp125 – 127)
- An FM receiver to tune the chosen IF (see, for example, reference [2], pp126 – 136 – there are many alternatives, see later)
- Various screened (metal) boxes, switches, plugs and sockets, controls and control knobs, cables, wire and general hardware – see later
- A 12V power supply at 1.5A maximum (including a scanner receiver), battery for portable use, mains for fixed station use.

So, how does it all work?

THE RECEIVER

A basic superheterodyne receiver for any frequency consists of two main parts, as shown in **Fig 1**. The first is a front-end converter consisting of a local oscillator (LO) and a mixer to produce a lower intermediate frequency (IF), the sum or difference of the LO frequency and the received frequency. The reason for converting the high-frequency received signal down to a lower frequency is because it is easier to amplify, filter and demodu-

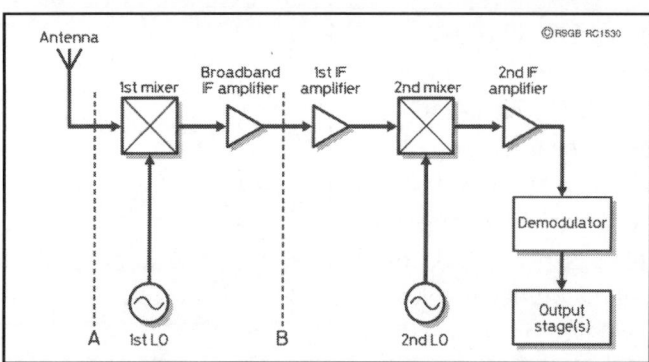

Fig 1. Block diagram of a basic double-conversion superheterodyne receiver. The front-end converter is between A and B, and the back-end is all to the right of B.

late. The second part is a back-end, consisting of IF amplifiers, filters, often a second oscillator/mixer (to convert signals to a second, even lower IF), a demodulator and an output stage.

A 10GHz receiver is no different. It would be very difficult for a beginner to make a microwave oscillator and mixer. Fortunately, there are ready-made oscillator / mixer units available in the form of surplus 'in-line' Doppler intruder alarm units. These are obtainable from most amateur radio rallies for £5 or £10. They consist of a Gunn device oscillator and a diode mixer mounted in-line inside a short length of (usually) waveguide 16 (WG16) which forms cavities (tuned circuits). WG16 is the standard waveguide size for 10GHz (3cm).

An aerial is connected to the open end of the waveguide and the received signal collected by the aerial travels down the waveguide to the mixer diode. LO input to the mixer is via a hole or slot in the end wall of the LO cavity. The IF output is taken from the mixer diode terminal on the Doppler module and connected to a suitable receiver back-end.

The back-end of the receiver can be almost anything you like, provided that it will tune to an IF somewhere between 10.7MHz (minimum) and, say, 50 or 60MHz, and has the necessary amplifiers, filters, demodulator and output stages. IFs below 10.7MHz can't be used because of LO noise, while IF signals above about 60MHz may be bypassed by the mixer diode decoupling capacitor built into the Doppler module. At one time a favourite choice was an FM broadcast band receiver covering 88 to 108MHz: these days there are so many FM stations in that band that it is almost impossible to avoid IF breakthrough.

Several choices can be made: a complete 10.7MHz back-end (see reference [2], pp12 6–136), a simple 50MHz receiver (see reference [2], pp74 – 86), or a scanner tuned to whatever IF seems to work best. The choice is yours.

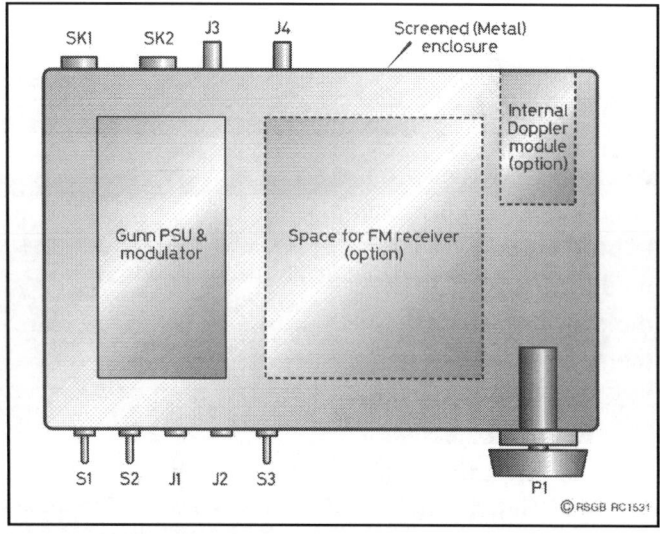

THE TRANSMITTER

Fig 2. A suggested layout for a base unit, extended to include a switchable internal Gunn.

The power output of the Doppler module Gunn oscillator is usually in the range 10 to 20mW. The mixer diode uses only a small proportion of this, leaving the remainder to escape from the open end of the Doppler module which of course is connected to the aerial. Since both the receiver input and the transmitter output (the open end of the Doppler module) are connected to the aerial, we don't need to switch the aerial between transmit and receive.

All we need in order to use the Gunn oscillator as a transmitter is some means of modulating it with speech or an audio tone.

By now, you can probably see that the heart of a simple 10GHz transceiver is the Gunn oscillator used both for receive and transmit. The Gunn oscillator needs a stabilised, variable, low-voltage bias power supply (typically 6 to 9V at about 150mA) to operate. If you want to know exactly how a Gunn device works, refer to reference [3]. Two factors determine the exact Gunn oscillator frequency – the size of the cavity (tuned circuit) in which it is mounted and the bias voltage which is applied.

The cavity can be tuned through several hundred megahertz by means of a metal screw in the cavity wall near to the Gunn device. This is a useful way of coarse tuning; inserting the screw into the cavity lowers the oscillator frequency, while withdrawing it raises the frequency.

Varying the Gunn bias voltage also changes the oscillator frequency. This is an effect known as frequency pushing. Lowering the voltage decreases the frequency, while raising it increases the frequency. The effective tuning range is limited to a few tens of megahertz, so that this is a very convenient way of 'fine tuning' the oscillator. A few millivolts of audio or tone modulation can be added to the bias voltage to produce high-quality FM modulation. The Gunn bias (power) supply therefore needs to be variable but, once set to a particular voltage, must be stable and free from noise. An unstable voltage or a noisy supply will result in an unstable and noisy oscillator! A very reliable Gunn supply with modulation facilities was described in reference [1], pp94 – 97. Use a 10-turn potentiometer and turns-counting dial for the 'fine-tune' control – this gives good bandspread and frequency-setting accuracy and is well worth the extra cost. The turns-counting dial can be calibrated in frequency to make frequency setting easier.

OTHER CONSIDERATIONS

Two other things need to be considered. Problem number one: coaxial feeder can't be used between the aerial and the Doppler module because the feeder losses at 10GHz would be astronomical! Problem number two: the IF output level of the mixer is very low so, again, losses in any connecting cable are not acceptable. What's to be done?

Both problems can be quite easily resolved by building the transceiver in two units. Mount the Doppler module and wide-band amplifier in a waterproof, screened (metal) box, attach the aerial directly to the same box and mount the whole of this 'microwave head' at the top of a mast for fixed station use – see reference [1], p94, or reference [2], p126, or on a tripod for portable use.

Build the Gunn power supply/modulator and all the controls (including the receiver back-end) into a base unit which can be used in the shack for fixed-station use or in the car for portable use.

Of course, if you want to work both fixed and portable, you will want to leave the mast-mounted microwave head at home and use a second, identical head for portable use. Build two masthead units or build another Doppler module into the base unit and use simple switching to change from the 'local' to the 'remote' microwave module, as suggested in **Fig 2**. S1 is the power on / off switch, S2 selects speech / tone, J1 is the mic jack, J2 the key jack, J3 coax to external receiver, J4 coax to mast-head unit, SK1 is for 12V in, SK2 for 12V to external receiver, P1 is the Gunn 'fine tune' control – 10-turn potentiometer / turns counter. The additional control S3 switches internal / external Gunn.

Built like this, one ordinary coaxial cable can now be used to connect the base unit to the remote unit, sending the modulated Gunn bias up to the remote microwave head and the amplified (broad-band) IF down from the remote microwave head to the base unit.

Full duplex operation was mentioned (see reference [2], p125). Unless the two stations share an identical IF, this will not be possible and it is probably the exception rather than the rule. A simple addition, shown in **Fig 3**, allows the G4KNZ Gunn PSU / modulator to be set for fixed-frequency transmit, while allowing receiver tuning over the whole range of frequency pushing. When everything appears to be working correctly, the last things to be done are to retune the Doppler Gunn oscillator from its usual ISM (Industrial, Scientific and Medical) frequency of 10,687MHz into the amateur band, somewhere between 10,370 and 10,400MHz, which is the current UK WBFM sub-band. To do this you'll need the help of someone with a 10GHz counter or high-Q wavemeter. If all else fails, try one of the Microwave Round Tables; dates and venues are announced in the 'Microwave' column in RadCom.

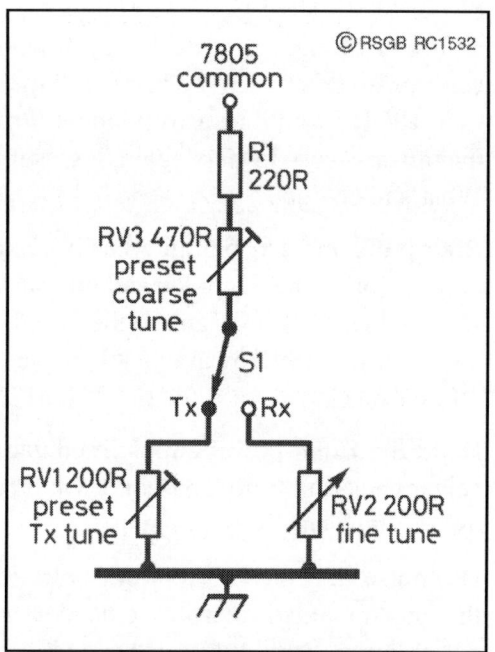

Setting up a Gunn oscillator is described in reference [2], pp134 – 135. Setting up the G4KNZ module, as modified in **Fig 2**, is very similar. You should have an oscillator tuning range of about ±10MHz. The receiver will respond to both images, that is, fosc ± FIF. If the IF is 10MHz, this means that the receiver will tune from 10,370MHz minus 10MHz to 10,390MHz plus 10MHz, ie from 10,360 to 10,400MHz. Using either image means that you can tune both the narrowband segment (10,368 to 10,370MHz) and most of the wideband segment (10,370 to 10,410MHz). You might be lucky enough to hear a beacon which will provide a very accurate frequency marker to check your calibration.

Fig 3. Modifications to the G4KNZ Gunn PSU / modulator to give a fixed transmit frequency and a tunable receiver.

CONCLUSION

It is possible to construct an entirely practical 10GHz WBFM transceiver using ready-made 10GHz Doppler oscillator / mixer modules, with some kind of FM receiver back-end. The overall cost of the whole project will depend on how much you like to experiment, how elaborate a transceiver you want to make, how carefully you shop around for parts and, to some extent, on how big a junk box you have! The point is that you do not have to be a skilled constructor or operator to get results.

Don't expect to work the world with simple, low-power 10GHz WBFM, but do expect to have a great deal of fun and gain useful experience and skills in trying to work your pals, either from home or portable (over much longer distances) using the 10mW or so transmitter power that typical, simple WBFM equipment produces.

Since there's plenty of bandwidth available, Doppler modules can be used for amateur fast-scan television (ATV) or very high-speed packet links – but that's another story. Think about it – microwaves have a lot to offer.

REFERENCES

[1]   Practical Transmitters for Novices, GW4HWR, RSGB.

[2]   Practical Receivers for Beginners, GW4HWR, RSGB.

[3]   Microwave Handbook, Vols 1, 2 and 3, ed G3PFR, RSGB.

*   The references above I believe are out of publication but a quick search of the internet and I found copies on ebay etc.or if you are like me you may have a copy in your shack – *Ed*

5.2   Screening – What Is It and Why Is It Important?

Screening or shielding, as it is sometimes known, can be very much a practical exercise when it becomes necessary to restrict a field or fields, close to their source, or alternatively to prevent a field or fields from reaching a sensitive point in a circuit. However, unless the underlying principles are applied, the outcome may not be successful. Textbooks are vague on the subject, so perhaps the following few 'rules of thumb' may go some way toward a better understanding.

As far as this feature is concerned, there are two fields, magnetic and electric. They can and do exist independently but, the instant one of them changes, an electromagnetic field is produced which is able to re-produce itself and propagate into the surrounding space (an electromagnetic wave), or possibly be constrained to travel along a transmission line of some description. For example, if a DC source is switched briefly into a transmission line of any length, the capacitance of the line causes a charging current to flow. This, along with the applied voltage, creates an electromagnetic wave which, willy-nilly, has to set off down the line.

Only changing or alternating magnetic or electric fields create electromagnetic waves. At low frequencies, eg at 50Hz (mains supply), the radiation is very small, as it is also at audio frequencies. The field surrounding the secondary of a mains transformer of a modern solid-state transceiver would be almost entirely magnetic, whereas the field surrounding the supply terminals of a valve linear power supply (2kV @ <1A) would be almost entirely electric.

MAGNETIC SCREENING

Now to the question of screening. There are basically two methods available:

(1) diverting the path of the field; and

(2) cancelling it out with an opposing field.

As an example of (1), consider **Fig 1**. Here the magnetic flux of

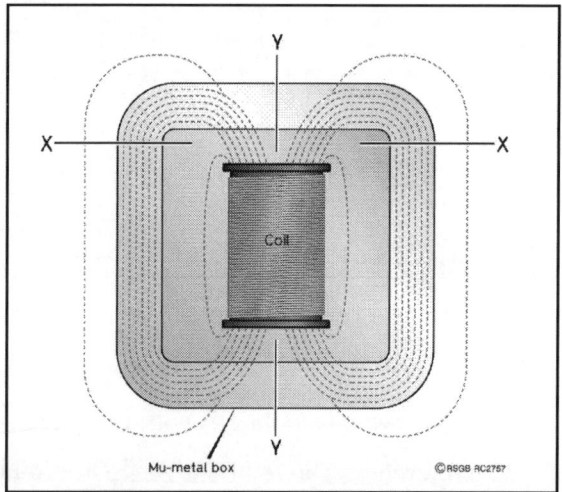

Fig 1. The high-permeability box concentrates the flux within the walls. Remember not to join the box along the line X-X.

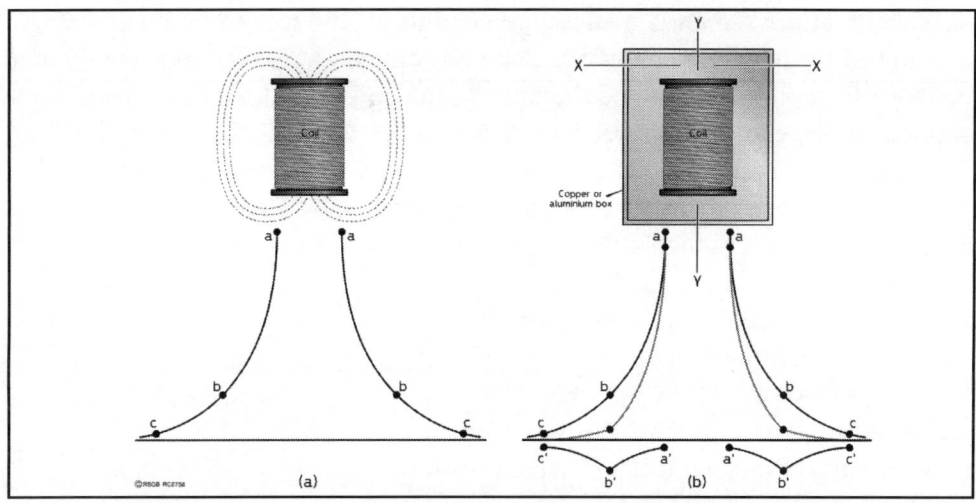

Fig 2(a). A simple coil and field. The flux density is shown by the graph being greatest at 'a' and least at 'c'. (b) The coil is now enclosed in a metal box. Below the line the graph shows the flux set up in the box; the letters 'a', 'b' and 'c' indicating the flux density. Note that it is the reverse of that in the coil. If we add the graphs above and below the horizontal axis, we derive the inner curve above the axis.

the coil finds an easy path through the high permeability of the enclosure, leaving only a very small residual field outside. It should be remembered that the lines depicting the path of the magnetic flux do not mean that the flux is made up of lines… it is spread throughout the space, though only very very thinly beyond the enclosure.

For the enclosure to be reasonably light in construction, the material used must have a very high permeability. It is here that an alloy called mu-metal is employed, though whether this is readily available nowadays is uncertain. At one time microphone transformers were all enclosed in mu-metal boxes, to prevent low and audio frequencies being picked up by the windings. Note that this method increases the inductance of the coil. A further very important point is to realise that the join in the box should not be made along the line X-X; to do this would insert a high resistance (reluctance) path to the flux. The join should be made along a line Y-Y instead. A mu-metal enclosure is effective from 0Hz to the higher audio frequencies, but as frequency increases the permeability of the alloy diminishes and the resistance increases, so method (2) is employed.

For method (2), copper or aluminium is used. Either material has a low resistance to current flow though the relative permeability is that of air, ie 1, hence it cannot offer a path for the magnetic flux. What it does do is act as a short-circuited secondary to the coil, and the current induced produces a field which tends to cancel out the field produced by the coil. This reduces the inductance of the coil but, provided that the screen is made large enough, the effect can be tolerated and/or allowed for in the design of the coil and screen. The screen diameter should be twice the coil diameter, and the ends of the coil should not come within one diameter of the ends of the screen. **Fig 2**(a) shows a coil with the normal flux lines depicting a magnetic field. The letters 'a', 'b', and 'c' indicate the falling-off of flux density as one moves away from the coil, the field being strongest at 'a' and much lower at 'c', though still assumed to be greater than wanted. We assume the current in the coil is alternating, therefore when the screen is placed around the coil, forming a closed circuit, currents will also be produced in it, though their direction will be opposite to those creating them (Lenz's Law). Provided that the screen resistance is very low, the currents produced in it will tend to reduce the flux at 'b' almost to zero, the small difference being that which is needed to create the reverse flux, 'b', see **Fig 2**(b). The letters 'a', 'b' and 'c' below the axis of the graph show the level of the flux present due to the screen at these points – note that they are in the reverse direction to those causing them and very slightly less in amplitude at 'b' and 'c' and very much less at 'a'. Note also that the flux at 'a' will only marginally affect the flux in the coil, therefore the inductance is not greatly reduced. It should be pointed out that the direction of the current in the screen is very different from that of the flux in **Fig 1**; it is in a continuous ring, parallel to the turns of the coil, hence the joint in this screen may be made in the direction X-X, but not in the direction Y-Y (which would prevent current flowing). As frequency is increased, the depth of field penetration into the screen becomes less, so relatively thin copper or aluminium sheet may be used, but for best results the boxes or enclosures must be watertight (the ideal), requiring joints to be lapped and special care taken with lids, covers, etc. *See* [1] and [2].

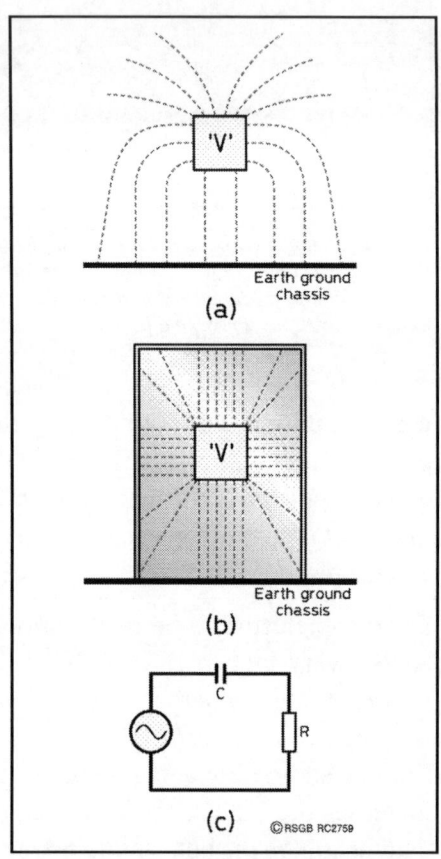

Fig 3(a). The point 'V' is at high potential and surrounded by an electric flux or field. (b) The field or flux is contained inside the screen. Every part of the screen is at earth potential, consequently the field or flux is no longer present outside. (c) The AC generator represents an HF or VHF source of voltage. Due to the presence of the screen, some capacitive coupling is certain. Should the screen be of poor conductivity, a potential difference will develop across it, giving some loss.

Incidentally, method (2) is completely useless for static or DC fields, and is not successful at low frequencies, therefore use mu-metal for DC and low frequency and copper or aluminium for RF screening.

ELECTRIC SCREENING

Perhaps you will have noticed that nothing has been said about 'earth' so far. This is because earthing has nothing to do with magnetic screening. However, it has everything to do with electric screening. It is the conductivity that is important. To screen a DC or LF electric field all you need to do is to enclose the field in an earthed can, box or enclosure.

Fig 3(a) represents a high voltage point, as may be found in a valve linear's power supply. Between this point and chassis/earth there will be an electric field, depicted by the dotted lines. This means that every point in space between 'V' and chassis/earth is at some potential (and hence not zero). If this field is enclosed in a conducting screen connected to chassis/earth, then the screen is considered to be at the potential of the chassis/earth, all points on the screen are at chassis/earth potential, so there is no longer an electric field outside the screen. This is true for non-varying or LF fields, but if the potential at 'V' is varying at a high frequency then there will be a capacitive current between 'V' and the screen, which could cause a potential difference across the screen, so some loss is possible. In practice it is advisable to have the screen as far away as possible from 'V', to avoid too much capacitive loading of the screened high voltage point.

It is not necessary to have a continuous metal screen, it may be made up of individual conductors in the form of a cage.

Although important, the joints need not have the perfection of those in a magnetic screen, so lids etc may be just a push fit. The degree of screening is measured by taking the ratio of the field strength before and after the screen is fitted. Screen effectiveness below 20dB, poor; from 20 to 80dB, average; 80 to 120dB, above average; above 120dB, cost problems.

REFERENCES

[1]   Practical RF Handbook, by Ian Hickman. Newnes, 1993. Good Appendix.

[2]   Circuit Designer's Companion, by Tim Williams. Newnes, 1993. Pages 248-252.

The prime purpose of a balun (a contraction of 'balanced-to-unbalanced') is to allow an unbalanced source to drive a balanced load or vice versa. Some types of balun will also yield an impedance transformation but this should be regarded as a secondary function.

BALANCED SYSTEM

Before getting into the details of baluns, it is necessary to understand just what is meant by a balanced load, and why feeding such a load from an unbalanced source can create problems.

Fig 1 shows a typical balanced load. The arrangement is symmetrical about the centre line. Each point on the left-hand side is mirrored by an equivalent point on the right-hand side, where the currents and voltages are equal in amplitude but opposite in phase.

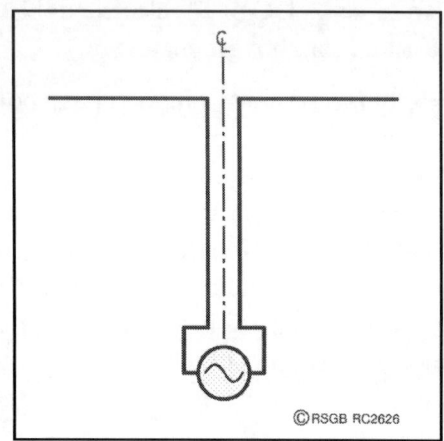

©RSGB RC2626

Fig 1. The standard dipole is electrically symmetrical about the centre line.

In the dipole itself, the currents in the two legs create fields which add together to generate the usual 'figure of eight' radiation pattern. The fields generated by each half of the feeder, though, cancel out each other so that there is no radiation from it.

FEEDING VIA COAX

Now consider the same dipole fed through a length of coaxial cable from a typical transmitter, as shown in **Fig 2**(a).

The current flowing in the inner conductor of the cable has only one destination, the left-hand leg of the dipole. That flowing in the outer of the cable, however, has two destinations – the right-hand leg of the dipole and back down the outside of the cable to ground.

Fig 2(b) shows a somewhat simplified representation of the various current paths, with I3 being that flowing back down the outside of the coaxial cable. The result of having a path for I3 is that a top-fed vertical aerial is, in effect, put in parallel with the right-hand leg of the dipole. This vertical aerial, will of course, radiate.

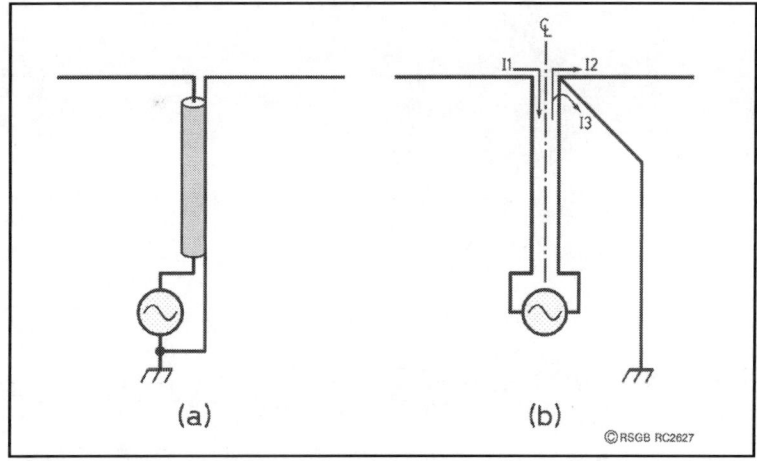

Fig 2. *If coax is used to feed a dipole, this generates a current path to earth, effectively resulting in a top-fed vertical antenna.*

(a) (b)

©RSGB RC2627

The amplitude of I3 is dependent upon the length of cable being used. If it is an odd multiple of a quarter-wave, the feed impedance of the effective vertical aerial is very high so I3 is low. Under these conditions the unbalancing effect on the dipole is insignificant. If, though, the cable is a multiple of a half-wave, the feed impedance is low and I3 is therefore high. As would be expected, the unbalancing effect is also high.

POTENTIAL PROBLEMS

Having explained the situation, the obvious question to be answered is: does it really matter? In order to answer this question, we need to consider two separate aspects: first, the effect on the aerial's ability to radiate; second, what might be called the system effects – such matters as EMC, VSWR measurement and so on.

Feeding a normal wire dipole via coax causes some change to the radiation pattern but this may not be particularly significant. There can in fact be some benefit, in that radiation from the vertical aerial represented by the outer of the coax fills in to some extent the nulls off the ends of the dipole. Should the dipole be part of an array, a Yagi beam for instance, the effects of feeding directly via coax can be quite significant. The forward gain will be reduced, as will the front-to-back ratio, and some side lobes will also appear. Similar problems will also occur with other forms of array based on balanced radiators such as the cubical quad.

Of the system effects, EMC is probably the one of greatest concern. In normal installations the coax will enter the shack and be in close proximity to house wiring. Radio-frequency energy radiated from the outer of the coax is highly likely to find its way into all the surrounding wiring and cause breakthrough problems.

Under normal circumstances, the VSWR on a transmission line is dependent only

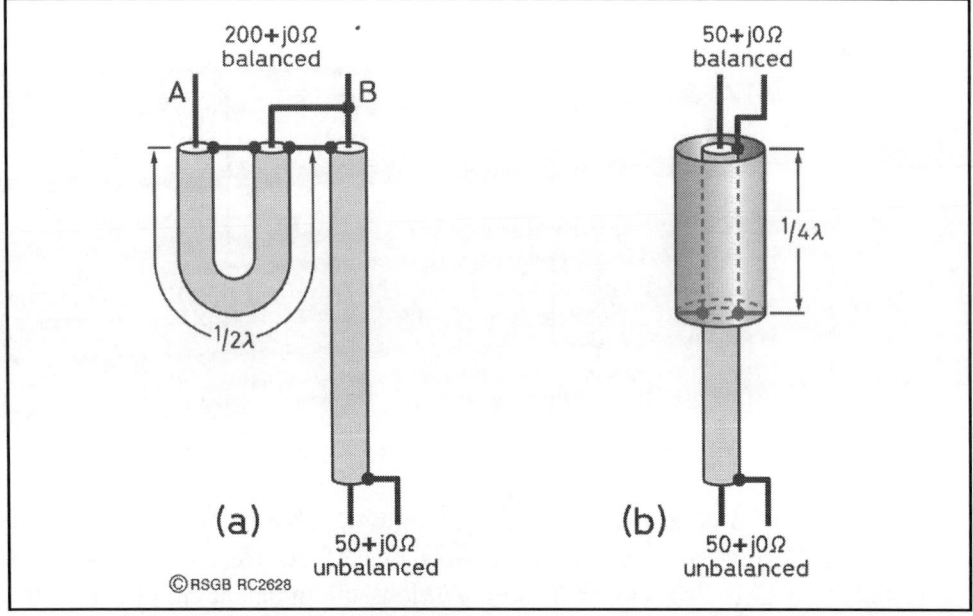

Fig 3. Narrow-band baluns can be made quite simply. A half-wave of coax, connected as in (a), generates a 180° phase shift. A quarter-wave sleeve, (b), effectively decouples the last part of the coax

upon the load impedance and the characteristic impedance of the line. Changing the line length should not cause the VSWR, as indicated by the usual VSWR meter, to change. Sometimes, though, it is found that the VSWR reading does change quite significantly with line length. This variation of reading is indicating that there is something odd about the setup. Feeding a balanced load via coax represents one of the possible oddities. **Fig 2**(b) shows that the actual load is made up of the dipole plus a top-fed vertical aerial. Also, as mentioned previously, the level of I3 is dependent upon the line length. In effect, then, the load impedance at the aerial terminals varies with line length. With VSWR being dependent upon this load impedance, varying the line length will cause the VSWR meter reading to change.

It is also possible that I3 flowing back down the outer of the coax will affect the operation of the VSWR meter, thus causing an additional source of confusion.

So, having established that feeding a balanced load via coax causes the radiation pattern of the aerial to change, the EMC threat to increase and the measurement of VSWR to become somewhat unpredictable, the next question is: can anything be done to improve the situation? Fortunately the answer is 'yes', and one of the solutions is to use a balun.

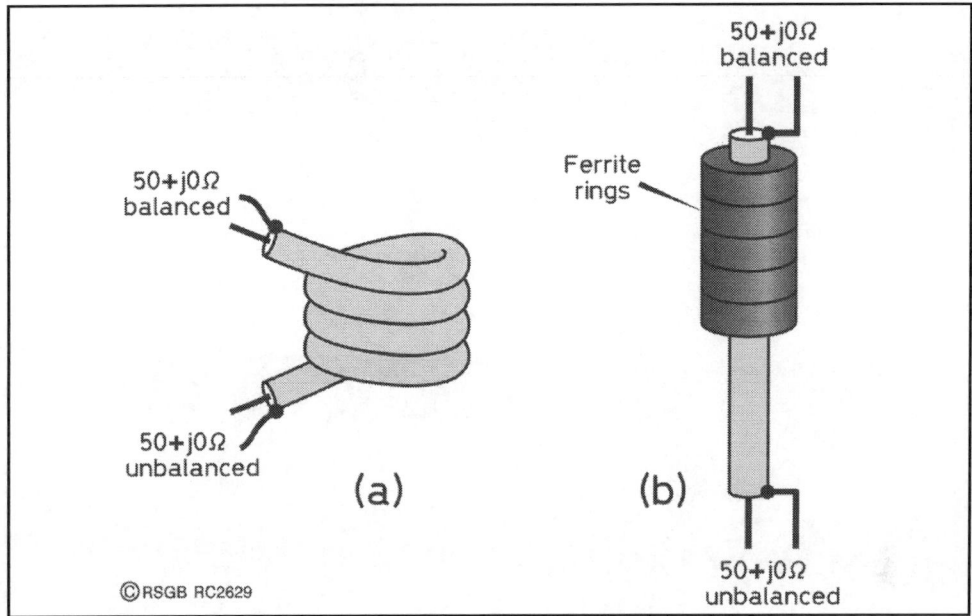

Fig 4. Two simple current-mode baluns. Both use inductance to reduce the current on the coax outer.

NARROW-BAND BALUNS

There are many types of balun, each having its advantages and disadvantages. Let's look first at some narrow-band arrangements – these being suitable for single-band aerials.

Fig 3(a) shows a very simple arrangement using an additional half-wavelength of coax. One feature of a half-wave line is that the voltage at the output is equal in amplitude to that at the input but with the phase reversed. With the arrangement shown, the voltages at A and B are equal in amplitude but of opposite phase. The voltage between A and B is twice that of the input. As a result, the load impedance for a 50Ω input impedance must be 200Ω. It should be noted that the aerial is not connected to the outer of the coax. If the inner of the coax does not have a DC path to earth, there can be problems with static build-up, particularly when there are electrical storms in the locality.

If a 1:1 impedance transformation is required, the arrangement shown in **Fig 3**(b) will suit the need. This uses a quarter-wave sleeve effectively to decouple the last section of the coax. Note that the top end of the sleeve needs to be well insulated and the bottom end connected to the outer of the coax. Although these two baluns have the virtue of simplicity, they are only really suitable for the higher frequen-

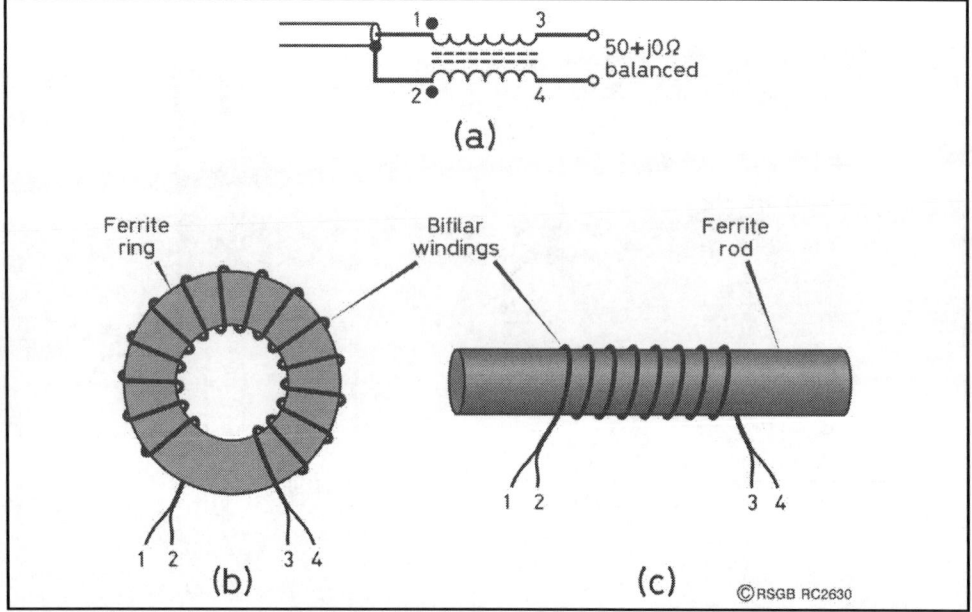

Fig 5. *A bifilar winding on a ferrite core, either a toroid or a rod, yields a simple current-mode balun. Note that the dots indicate the starting points of each winding.*

cies owing to the lengths of line needed to make them.

BROAD-BAND BALUNS

These can be one of two basic types: those which force the currents in the two halves of the aerial to be equal in amplitude but of opposite phase and those which force the voltages to have this relationship. If the aerial is truly balanced, both achieve the same effect. One problem with wire aerials at the lower frequencies is that it is difficult for many reasons to achieve a fully balanced arrangement. The current balun ensures that in such cases the currents in both conductors of the feeder are equal in amplitude.

Fig 4 shows two simple types of current mode balun, both of which provide a 1:1 impedance ratio. These work on the principle of providing an impedance to restrict the flow of an out-of-balance current. In the case of **Fig 4**(a) this is achieved by coiling up the coax near to the feed-point of the aerial. The coil has no effect on the normal signal flowing up the coax but looks like an inductance to any current trying to return via the outer. The same effect is achieved in **Fig 4**(b) by threading ferrite rings over the coax.

One difficulty with the arrangement shown in **Fig 4**(a) is that it can be difficult

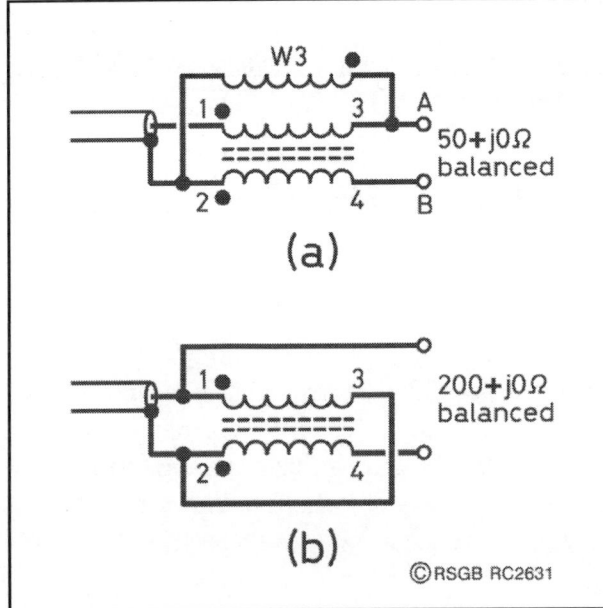

Fig 6 (a). Adding a third winding, W3, to the simple current-mode balun yields a 1:1 voltage-mode balun. (b) Connecting the two bifilar windings in a different way gives a 4:1 impedance step-up design.

to make it work effectively over a wide frequency range – say, 3.5 to 30MHz. Providing sufficient turns to cope with 3.5MHz is likely to result in the inter-winding capacitance being too high for effective operation at 30MHz. The situation can be improved by winding the turns on a ferrite rod or ring.

A current-mode balun can also be constructed by using a bifilar winding on a toroid or ferrite rod, as shown in **Fig 5**. It must be understood that these bifilar windings act like transmission lines, which can limit the performance of the arrangements in some circumstances and yield rather erratic results. Ruthroff [1] advocated the addition of a third winding to the simple bifilar winding of **Fig 5**(a) to yield the arrangement shown in **Fig 6**(a). Although this third winding overcomes some of the problems of the two-winding arrangement, it has the effect of turning the balun into a voltage-mode device. Windings 1 – 3 and W3 act like an auto-transformer, so that the voltage at point A is half that of the input. The voltage at point B will also be half that of the input but with a phase reversal. This arrangement tends to be regarded as the 'standard' for 1:1 impedance ratio baluns.

The simplest form of voltage-mode balun, albeit with a 4:1 impedance step-up, uses the same basic arrangement as **Fig 5**(a), but with the windings connected in a different way, as shown in **Fig 6**(b). Construction is as for the examples in **Fig 5**(b) and **Fig 5**(c).

There are two (often conflicting) criteria associated with the design of voltage-mode baluns. First, the inductive reactance of the windings should be high; second, the leakage reactance should be low compared with the load impedance in each case. The first usually determines the low-frequency limit of operation whilst the second determines the high-frequency limit.

CONCLUSIONS

This has been a fairly brief introduction into the subject and has (quite deliberately) begged the question of which design is 'best'.

The reason for this omission is that the use of baluns tends to result in compromises having to be made – what works well in one application may be a total failure in others. The sensible approach is to try a few different ideas and select that which gives the best performance in your particular set-up. Fortunately, the components used are reasonably inexpensive and can easily be recycled for the different arrangements.

For those wanting to have a go, references [2 – 7] list articles and books containing more details of the different arrangements. You might be bemused by the fact that some authors will be enthusiastic about a particular arrangement whilst others regard it with horror. Take due note of any objections to the various designs but do not let these put you off trying them.

REFERENCES

[1]'Some broad-band transformers', C L Ruthroff, Proc IRE, Vol 47, August 1959.

[2]'Balance to unbalance transformers', Ian White, G3SEK, Radio Communication December 1989 (highly recommended reading).

[3]Radio Communication Handbook, RSGB.

[4]ARRL Handbook, ARRL.

[5]HF Aerials for All Locations, Les Moxon, G6XN, RSGB.

[6]Backyard Aerials, Peter Dodd, G3LDO, RSGB, 2000.

[7]Transmission Line Transformers, ARRL.

[8]Reflections, Transmission Lines and Aerials, ARRL.

Aerial Maintenance

Spring is the ideal time to do something about those aerials that have suffered during the extremes of the winter months. This work involves maintenance of existing aerials and the installation of better ones.

AERIAL UPKEEP

We all know that a length of wire will radiate but a length of wire will radiate even better if it is in good condition. Most aerial losses are caused by corrosion, which increases the resistive losses in the aerial. When maintenance is carried out on an aerial system, all mechanical joints should be dismantled and corrosion removed. They can then be coated with grease before reassembly. Soldered joints should be inspected and remade if they look suspect. Insulators, particularly at high voltage points (such as at the ends of dipoles or long wire aerials), should be cleaned.

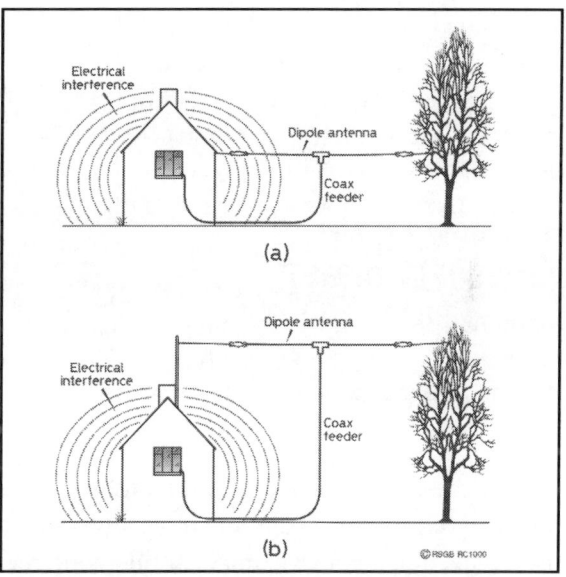

Fig 1. (a) Dipole aerial at a low height, showing it in the strong electrical interference field; (b) aerial height is increased, thus reducing the electrical interference and increasing the signal.

Check that the transmission lines (coaxial cables or twin feeders) are in good condition and that the connectors are free from corrosion. They, too, should be coated in grease after cleaning and before re-assembly.

IMPROVING THE AERIAL

In general, the most practical way of improving the performance of an existing aerial is to erect it as high as is practicable at your location. An aerial that is low and close to the house is also close to the electrical QRM which envelopes our dwellings as shown in **Fig 1**(a). If the aerial can be raised it will be further from this interference, which will enable us to hear stations that would otherwise be hidden in the electrical noise; see **Fig 1**(b).

The aerial will also be further away from TV sets and other domestic electronic equipment and the chances of TVI are reduced. Also, when the height of the aerial is increased, the radiation pattern favours DX stations to a greater degree (because of the lower angle of radiation).

A worthwhile improvement can often be made with a modest increase in height, such as moving one end of the aerial from the eves of the house to the chimney or connecting it to a higher branch of a tree.

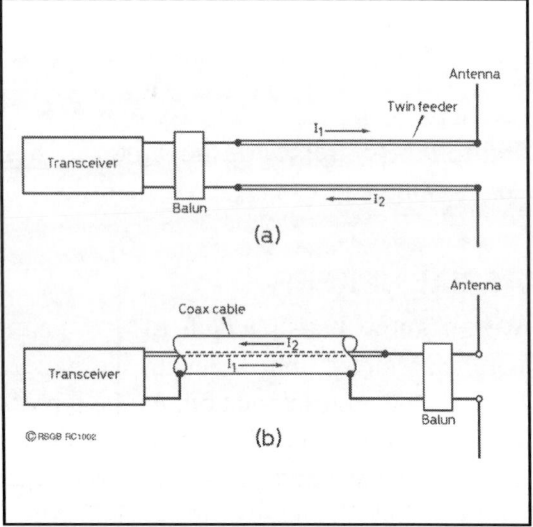

Fig 2. Showing the field-cancelling effects of current in (a) twin feeder and (b) coaxial cable.

TRANSMISSION LINES (FEEDERS)

Although the aerial is installed as high as possible, the transceiver is installed at a convenient place indoors, where it is readily accessible and out of the weather.

For the best signal-to-noise ratio, and for avoiding TVI, it is better if the connector between the aerial and the transceiver does not radiate or receive signals. This is achieved using a transmission line, which can comprise coaxial cable ('coax') or twin-line feeder.

The twin-feeder form of transmission line comprises two conductors placed parallel and close together as shown in **Fig 2**(a). The current flowing in the two conductors travels in opposite directions; in other words they are 180° out of phase with each other. If the two currents also have equal amplitudes, the electromagnetic field generated by each conductor will cancel that generated by the other, and the line will not radiate or receive radio energy.

With coax (**Fig 2**(b)), the current flowing in the outer conductor does so on the inner surface so that radiated or received RF energy is cancelled, just the same as in the twin-line feeder. In addition, the outer conductor of coax acts as a shield, confining the RF energy within the line. Because of its construction, coaxial cable is said to be unbalanced.

If the currents flowing in the two lines of the feeder are not equal in amplitude or not exactly 180° out of phase, the line will radiate or receive RF energy.

Centre-fed dipoles and loops are balanced, meaning that they are electrically symmetrical with respect to the feed-point (where the feeder is connected to the aerial). A balanced aerial should be fed with a balanced feeder system to preserve this electrical symmetry with respect to ground. However, the aerial connector on the back of the transceiver is for coaxial cable, and is thus unbalanced. Some method of method of connecting the transmission line to the aerial without up-setting the symmetry of the aerial itself is required. A device for converting a balanced circuit to an unbalanced circuit is a balun (a contraction of 'balanced to unbalanced'); see **Fig 3**.

Coaxial cable has the advantage of being very practical for most amateur radio installations. Because of the excellent shielding afforded by its outer screening, coax can be run up a metal tower or taped together with numerous other cables with virtually no interaction. At the top of a tower, coax can be used with a ro-tating beam without shorting or twisting conductor problems. Coax can even be buried underground in plastic tubing.

The disadvantage of it is that it normally requires some matching unit or balun at the aerial. Also, precautions have to be taken to ensure that the connection is ab-solutely watertight. If water gets into the tube-like structure of coax then the cable is ruined. Coaxial cable is relatively heavy compared with a single copper wire. This means that there is a fair amount of mechanical stress on the coax-to-aerial connection of a centre-fed dipole, espe-cially when using heavy-duty coax.

Open-wire feeder must be kept away from metal objects by several times the spacing between its conductors. Despite this mechanical difficulty, there are often compelling reasons for using this type of feeder. One of these is the low loss in-curred when using twin-line feeder in a multi-band aerial. In addition, twin feeder is far cheaper than coaxial cable and it is much lighter.

An excellent and cheap insulator for HF aerials is a 1m long length of heavy mono-filament fishing line or strimmer cord, but learn to tie a bowline or fisherman's knot otherwise it will easily come undone!

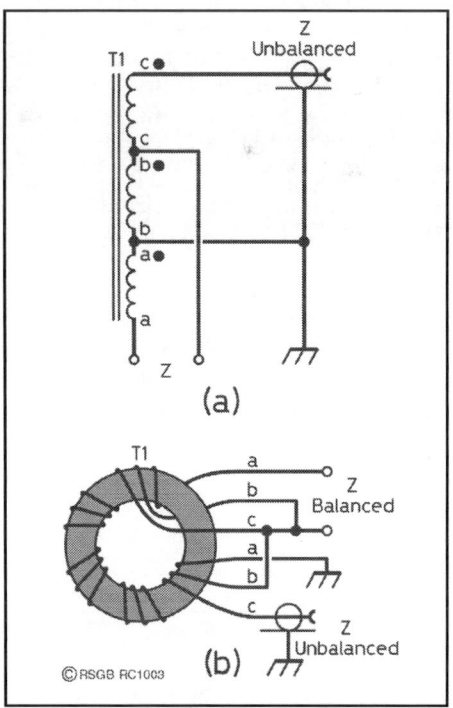

Fig 3. *(a) Circuit diagram of a 1:1 balun; (b) its physical construction.*

Software Defined Radio

By Andrew Barron, ZL3DW

Everyone is talking about software defined radio (SDR) but is SDR right for you? *Software Defined Radio* sets out to explain the basics without getting to technical and is written to help you too get the most out of your SDR. It will even help you decide what to buy.

Written by New Zealand based and acknowledged SDR expert Andrew Barron, ZL3DW, *Software Defined Radio* covers a huge range of material. The use of SDR by radio amateurs is growing rapidly in popularity as they become aware of the great features and performance on offer. Not only does this book cover how SDR works there are details the different types of software that are available, what is different about them and even what is better. There is a wealth of useful information included and even guides to what to look for when you are buying equipment. There are guides to using SDR with CW, Digital Modes, Contesting, EME, Microwaves, Satellites and much more. You will find information on over 60 SDR radios that you can buy today featuring leading brands such as FlexRadio, Elecraft, Anan, Expert, Elad, Icom, WiNRADiO, SDRplay, FUNcube and many more.

Software Defined Radio is intended for radio amateurs, short wave listeners or anyone interested in radio technology. If you are interested in the technology of what was once, the domain of a few dedicated hackers and experimenters, the future of this exciting and fast developing area of radio or simply want to buy a SDR radio, this book is thoroughly recommended reading.

Size 174x240mm, 304 pages
ISBN: 9781 9101 9349 5

Only £12.99 plus p&p

Radio Society of Great Britain www.rsgbshop.org
3 Abbey Court, Priory Business Park, Bedford, MK44 3WH. Tel: 01234 832 700 Fax: 01234 831 496